DIVERSITY IN TECHNOLOGY EDUCATION

Editor
Betty L. Rider, PhD, DTE
President, RidercoM, Inc.

47th Yearbook, 1998

Council on Technology
Teacher Education

GLENCOE

McGraw-Hill

New York, New York   Columbus, Ohio   Mission Hills, California   Peoria, Illinois

A Division of The **McGraw·Hill** Companies

Copyright © 1998 by the Council on Technology Teacher Education. All rights reserved. Except as permitted under the United States Copyright Act, no part of this publication may be reproduced or distributed in any form or by any means, or stored in a database or retrieval system, without prior written permission from the publisher.

Send all inquiries to:
Glencoe/McGraw-Hill
3008 W. Willow Knolls Drive
Peoria, IL 61614

ISBN 0-02-831274-0

Printed in the United States of America.

1 2 3 4 5 6 7 8 9 10   026   02 01 00 99 98

Orders and requests for information about cost and availability of yearbooks should be directed to Glencoe/McGraw-Hill's Order Department, 1-800-334-7344.

Requests to quote portions of yearbooks should be addressed to the Secretary, Council on Technology Education, in care of Glencoe McGraw-Hill at the above address, for forwarding to the current Secretary.

This publication is available in microform from

UMI
300 North Zeeb Road
Dept. P.R.
Ann Arbor, MI 48106
Phone: 1-800-521-0600, extension 2888

FOREWORD

Jay Cummings
President, National Association of State Directors of the
Vocational Technical Education Consortium

The confluence of change and diversity in an educational environment that supports the status quo can present a daunting challenge. The more persuasive arguments that herald the realities that must be encountered successfully, such as "change is the only constant" and "there is strength in diversity," seem to fall on deaf ears when people who are vested in the current system are confronted. Less than five years ago, forecasters and futurists predicted that the changing demographics would lead to a workforce composed increasingly of ethnic minorities and women as we move rapidly toward the millennium.

Instead of implementing the social and educational adaptations needed for a diverse workforce, including attitudinal adjustments to accommodate and capitalize on change and diversity, many otherwise rational people have pursued strategies of avoidance and denial. These strategies, when implemented and championed in public policy forums, cause our institutions, programs, and people to come to grips with the truism: "Anyone who does not learn from the lessons of history is doomed to repeat them."

Issues of representation, inclusion, access, and equity are at the forefront of the agendas for the current leadership in technology education, and any leader who is in a mode of denial or avoidance is "asleep at the wheel." The theme of the 47th yearbook, *Diversity in Technology Education*, goes straight to the heart of the matter for a profession that has not embraced women and ethnic minorities openly and willingly.

From a beginning that covers gender and ethnicity, the editor acknowledges that the book is a journey, not a destination. Other sage advice that the leadership in technology education might heed follows: "A thousand-mile journey begins with the first step." The Council on Technology Teacher Education (CTTE) has taken the first step through this publication, and that is encouraging. However, the scope of what diversity embraces in the broader social context augurs for consideration of disabilities, age, geography, and language.

In a 1991 publication entitled *America and the New Economy*, Anthony Carnevale outlined the nation's challenge and opportunity:

> Plain vanilla isn't good enough anymore. Variety and customization of goods and services have become key competitive principles. Consumers' tastes have diversified because of a flammable mix of economics and demography at home and abroad. Increasing economic wealth contributes to diversity in demand in two ways. First, it changes what people want. Second, growing wealth gives economic voice to underlying ethnic, geographic, cultural, religious, and gender differences that were there all along. (p. 20)

The book is sufficient as a tool to raise awareness and provoke thought. Whether this awareness and thought can facilitate appropriate action to meet the challenges facing technology education is the significant question before us.

Technology educators must address realities in equity and access as well when discussions of diversity in classrooms related both to teachers and students are undertaken. In some communities where students from economically, socially, linguistically, and culturally diverse families reside, there is an additional neglected dichotomy related to the "haves" and "have nots." In some quarters, the move toward technology is viewed as an attempt to maintain the status quo between the "haves" and the "have nots." To make an appropriate response, technology educators must take full account of the current disparities in services and student academic performance by underscoring how technology can be used for excellence and equity in achievement.

By choosing to examine diversity issues, CTTE, the editor, and the volunteer authors have welcomed the potential for change in the profession and the instructional programs that can make a tremendous difference in how technology educators respond to diversity. In the emerging discussions that this book is sure to generate, the possibility is great that technology educators engaged in conversation about diversity and change may reach the following conclusions:

> First Educator: If it is to be, it's up to me.
>
> Second Educator: If this is true, it's also up to you.

REFERENCES

Carnevale, A. (1991). *America and the new economy.* Washington, DC: American Society for Training and Development.

YEARBOOK PLANNING COMMITTEE

Terms Expiring 1998
 Stanley A. Komacek
  California University of Pennsylvania
 Brenda L. Wey
  Appalachian State University

Terms Expiring 1999
 Michael K. Daugherty
  Illinois State University
 Anthony F. Gilberti
  St. Cloud State University

Terms Expiring 2000
 Rodney L. Custer
  University of Missouri—Columbia
 Michael L. Scott
  The Ohio State University

Terms Expiring 2001
 Anthony E. Schwaller
  St. Cloud State University
 Donna K. Trautman
  Bowling Green State University

Terms Expiring 2002
 Karen F. Zuga
  The Ohio State University
 Gene Martin
  Southwest Texas State University

OFFICERS OF THE COUNCIL

President
John M. Ritz
Old Dominion University
4600 Hampton Blvd.
Norfolk, VA 23529-0498

Vice-President
Mark E. Sanders
VIP & State University
Technology Education
Blacksburg, VA 24061-0432

Secretary
W. Tad Foster
Central Connecticut State University
1615 Stanley St.
New Britain, CT 06050

Treasurer
Charles A. Pinder
Northern Kentucky University
Louie B. Nunn Dr.
Highland Heights, KY 41099

Past President
Everett N. Israel
Eastern Michigan University
Department of Industrial Technology
Ypsilanti, MI 48197

YEARBOOK PROPOSALS

Each year at the ITEA International Conference, the CTTE Yearbook committee reviews the progress of yearbooks in preparation and evaluates proposals for additional yearbooks. Any member is welcome to submit a yearbook proposal, which should be written in sufficient detail for the committee to be able to understand the proposed substance and format. Fifteen copies of the proposal should be sent to the committee chairperson by February 1 of the year in which the conference is held. Below are the criteria employed by the committee in making yearbook selections.

CTTE Yearbook Committee

CTTE Yearbook Guidelines

A. Purpose:
The CTTE Yearbook Series is intended as a vehicle for communicating education subject matter in a structured, formal series that does not duplicate commercial textbook publishing activities.

B. Yearbook topic selection criteria:
An appropriate yearbook topic should
1. Make a direct contribution to the understanding and improvement of technology teacher education.
2. Add to the accumulated body of knowledge in the field.
3. Not duplicate publishing activities of commercial publishers or other professional groups.
4. Provide a balanced view of the theme and not promote a single individual's or institution's philosophy or practices.
5. Actively seek to upgrade and modernize professional practice in technology teacher education.
6. Lend itself to team authorship as opposed to single authorship.

Proper yearbook themes *may* also be structured to
1. Discuss and critique points of view that have gained a degree of acceptance by the profession.
2. Raise controversial questions in an effort to obtain a national hearing.
3. Consider and evaluate a variety of seemingly conflicting trends and statements emanating from several sources.

C. The yearbook proposal:
1. The Yearbook Proposal should provide adequate detail for the Yearbook Planning Committee to evaluate its merits.
2. The Yearbook Proposal should include
   a. An introduction to the topic
   b. A listing of chapter titles
   c. A brief description of the content or purpose of each chapter
   d. A tentative list of authors for the various chapters
   e. An estimate of the length of each chapter

PREVIOUSLY PUBLISHED YEARBOOKS

*1. Inventory Analysis of Industrial Arts Teacher Education Facilities, Personnel and Programs, 1952.
*2. Who's Who in Industrial Arts Teacher Education, 1953.
*3. Some Components of Current Leadership: Techniques of Selection and Guidance of Graduate Students; An Analysis of Textbook Emphases; 1954, three studies.
*4. Superior Practices in Industrial Arts Teacher Education, 1955.
*5. Problems and Issues in Industrial Arts Teacher Education, 1956.
*6. A Sourcebook of Reading in Education for Use in Industrial Arts and Industrial Arts Teacher Education, 1957.
*7. The Accreditation of Industrial Arts Teacher Education, 1958.
*8. Planning Industrial Arts Facilities, 1959. Ralph K. Nair, ed.
*9. Research in Industrial Arts Education, 1960. Raymond Van Tassel, ed.
*10. Graduate Study in Industrial Arts, 1961. R. P. Norman and R. C. Bohn, eds.
*11. Essentials of Preservice Preparation, 1962. Donald G. Lux, ed.
*12. Action and Thought in Industrial Arts Education, 1963. E. A. T. Svendsen, ed.
*13. Classroom Research in Industrial Arts, 1964. Charles B. Porter, ed.
*14. Approaches and Procedures in Industrial Arts, 1965. G. S. Wall, ed.
*15. Status of Research in Industrial Arts, 1966. John D. Rowlett, ed.
*16. Evaluation Guidelines for Contemporary Industrial Arts Programs, 1967. Lloyd P. Nelson and William T. Sargent, eds.
*17. A Historical Perspective of Industry, 1968. Joseph F. Luetkemeyer Jr., ed.
*18. Industrial Technology Education, 1969. C. Thomas Dean and N. A. Hauer, eds.; Who's Who in Industrial Arts Teacher Education, 1969. John M. Pollock and Charles A. Bunten, eds.
*19. Industrial Arts for Disadvantaged Youth, 1970. Ralph O. Gallington, ed.
*20. Components of Teacher Education, 1971. W. E. Ray and J. Streichler, eds.
*21. Industrial Arts for the Early Adolescent, 1972. Daniel J. Householder, ed.
*22. Industrial Arts in Senior High Schools, 1973. Rutherford E. Lockette, ed.
*23. Industrial Arts for the Elementary School, 1974. Robert G. Thrower and Robert D. Weber, eds.
*24. A Guide to the Planning of Industrial Arts Facilities, 1975. D. E. Moon, ed.
*25. Future Alternatives for Industrial Arts, 1976. Lee H. Smalley, ed.
*26. Competency-Based Industrial Arts Teacher Education, 1977. Jack C. Brueckman and Stanley E. Brooks, eds.
*27. Industrial Arts in the Open Access Curriculum, 1978. L. D. Anderson, ed.
*28. Industrial Arts Education: Retrospect, Prospect, 1979. G. Eugene Martin, ed.
*29. Technology and Society: Interfaces with Industrial Arts, 1980. Herbert A. Anderson and M. James Benson, eds.
*30. An Interpretive History of Industrial Arts, 1981. Richard Barella and Thomas Wright, eds.
*31. The Contributions of Industrial Arts to Selected Areas of Education, 1982. Donald Maley and Kendall N. Starkweather, eds.
*32. The Dynamics of Creative Leadership for Industrial Arts Education, 1983. Robert E. Wenig and John I. Mathews, eds.
*33. Affective Learning in Industrial Arts, 1984. Gerald L. Jennings, ed.
*34. Perceptual and Psychomotor Learning in Industrial Arts Education, 1985. John M. Shemick, ed.
*35. Implementing Technology Education, 1986. Ronald E. Jones and John R. Wright, eds.
*36. Conducting Technical Research, 1987. Everett N. Israel and R. Thomas Wright, eds.
*37. Instructional Strategies for Technology Education, 1988. William H. Kemp and Anthony E. Schwaller, eds.
38. Technology Student Organizations, 1989. M. Roger Betts and Arvid W. Van Dyke, eds.
*39. Communication in Technology Education, 1990. Jane A. Liedtke, ed.
40. Technological Literacy, 1991. Michael J. Dyrenfurth and Michael R. Kozak, eds.
41. Transportation in Technology Education, 1992. John R. Wright and Stanley Komacek, eds.
42. Manufacturing in Technology Education, 1993. Richard D. Seymour and Ray L. Shackelford, eds.
43. Construction in Technology Education, 1994. Jack W. Wescott and Richard M. Henak, eds.
44. Foundations of Technology Education, 1995. G. Eugene Martin, ed.
45. Technology and the Quality of Life, 1996. Rodney L. Custer and A. Emerson Wiens, eds.
46. Elementary School Technology Education, 1997. James J. Kirkwood and Patrick N. Foster, eds.
47. Diversity in Technology Education, 1998. Betty L. Rider, ed.

*Out-of-print yearbooks can be obtained in microfilm and in Xerox copies. For information on price and delivery, write to Xerox University Microfilms, 300 North Zeeb Road, Ann Arbor, Michigan, 48106.

CONTENTS

Foreword iii
Yearbook Planning Committee vi
Officers of the Council vii
Yearbook Proposals viii
Previously Published Yearbooks ix
Preface xiii
Acknowledgments xv

─◇Chapter 1

Society, Diversity, and Technology Education 1
Donna K. Trautman
Bowling Green State University
Bowling Green, OH

SECTION I: HISTORICAL INFLUENCES OF UNDERREPRESENTED GROUPS IN TECHNOLOGY EDUCATION

─◇Chapter 2

A Historical View of Women's Roles in Technology Education 13
Karen F. Zuga
The Ohio State University
Columbus, OH

─◇Chapter 3

Contributions of African-Americans to Technology Education 37
Michael L. Scott
The Ohio State University
Columbus, Ohio

Keith V. Johnson
East Tennessee State University
Johnson City, TN

SECTION II: UNDERREPRESENTED GROUPS AS TECHNOLOGY STUDENTS AND EDUCATORS

—◇Chapter 4
Women as Technology Educators 57
Colleen E. Hill
California State University, Long Beach
Long Beach, CA

—◇Chapter 5
Minority Students 77
Elazer J. Barnette
North Carolina A & T University
Greensboro, NC

SECTION III: INCREASING AND SUPPORTING DIVERSITY

—◇Chapter 6
Reading, Writing, and Technology 97
Karen Coale Tracey
Central Connecticut State University
New Britain, CT

—◇Chapter 7
Mentors for Women in Technology 123
Daniel L. Householder
Texas A & M University
College Station, TX

—◇**Chapter 8**

Effective Leadership for All 139
  Elizabeth Smith
  Pensacola Junior College
  Pensacola, FL

—◇**Chapter 9**

Environmental and Climate Challenges in
Technology Education 149
  Jane A. Liedtke
  Illinois State University
  Normal, IL

SUMMARY

—◇**Chapter 10**

Diversity in Technology Education 177
  Janet L. Robb
  Gable Social Mobilization Campaign
  Zomba, Malawi

PREFACE

This book is not finished. There were those who claimed it would not get done, and they were right. There are so many issues that are *not* explored in this book that it should perhaps be called *Diversity in Technology Education: The Beginning of a Journey*. Our profession does not own the issues explored in this book. Issues related to diversity affect all areas of our lives and the world beyond where we live and work.

My journey began as a suburban Chicago junior high student, when I walked by the woodshop and thought that it looked like an interesting place. I was told that because I was female, I was not allowed to take industrial arts but had to take home economics. Four years later, during my junior year of high school, that situation changed when my brother, Bill, and his friends took a cooking class. My girlfriends and I went to the principal and demanded the right to take industrial arts classes. I said, "Look, Dad, you let Bill into home economics classes, so you have to let us into industrial arts classes." I'm not sure if it was the effects from Title IX or just keeping the peace at home, but we were enrolled. Roger Stieglitz was our woodworking teacher; taking the class was such a great experience that I re-enrolled in my senior year. Never was gender an issue in that classroom; the expectations were the same for all students.

As an undergraduate technology education major at Eastern Illinois University (EIU), our professors made us work hard and had high expectations for not only the quantity, but the quality of what we produced. Several women were enrolled in the program at that time, and gender was not an issue with my professors.

In my first teaching job in Wheeling, IL, "home arts"—a semester of home economics and a semester of industrial arts—was required of all students. Because of the size of the school, there were three technology teachers: two male, one female; two White, one African-American. Gender and race were not issues for the students or the staff.

As a master's student at EIU, I, along with all the other students, worked hard. Usually I was the only woman in the class, but the expectations of me were no different than of the men, whether they were White, Asian, African, Sri Lankan, or African-American.

Moving to another state to teach in a middle school created some situations where my gender was problematic for others. The vocational director was heard to have said upon my being hired, "You hired a *what* to teach industrial arts?" At my first regional meeting, one person suggested that I become the organization's secretary, as that was "women's work." At my first state conference, a participant asked, "Whose spouse are you?" However, in a positive light, I had more parents attend conference night than any teachers of the other elective subjects. I also served as a role model for female students; enrollment increased from 5 to 22 females over 2 enrollment cycles. At Texas A&M University, where I was the first woman technology educator to graduate with a doctorate in industrial education, I was again expected to meet the same criteria as my male counterparts.

Each time I graduated I was proud of my accomplishments; I was a technology education professional—no different than my male colleagues. That was my premise for functioning as a professional on the job as well as through service to my professional organizations. But I am different than my male colleagues. Although I brought similar skills and knowledge to the table, my perspective was different because of my gender. The same is true for persons from different cultures, who can provide diverse perspectives.

You might ask, So what? Why change? Who cares? or even What's in it for me? I hope that this yearbook helps to answer these questions. Demographically, the world is changing around us. If for no other reason than our survival as academicians, we should be proactive to ensure that our profession reflects the constituents that we serve.

ACKNOWLEDGMENTS

The Council on Technology Teacher Education (CTTE) is to be commended for initiating and sustaining this series of yearbooks. Over time the yearbooks have made a significant contribution to our profession. I am pleased to have had the opportunity to serve as the editor of the 47th Yearbook, *Diversity in Technology Education*.

Glencoe/McGraw-Hill and its predecessor, McKnight & McKnight Publishing Company, have been unwavering supporters of technology education. They are ever faithful in serving our profession and publishing the yearbook series. Their staff continues to be helpful, especially Trudy Muller and Stacey Stull, who provided an abundance of support, guidance, and advice.

There is now another reason to applaud Glencoe/McGraw-Hill. This year they have added a CD-ROM to accompany the book. I appreciate the support from Dave Whiting and Cathy Scruggs on this new venture. The CD-ROM is the brain child of Dr. Donna K. Trautman, who also provided the technical skill and knowledge to produce the master for production. The CD-ROM programmer, Gretchen Weller, a senior visual communication technology student, has done an outstanding job.

The genesis of this book came during a meeting in Normal, IL, in February 1994. The Technical Foundation of America awarded a grant to Jane Liedtke to bring together women leaders to develop grants and activities to support women in technology education. Participants, in addition to Jane and me, were Debra Bozarth and Drs. Colleen Hill, Donna Trautman, Rosanne White, LaVerne Young-Hawkins, and Karen Zuga. Many times the conversation turned to situations we had run into as women in a profession that is predominantly male. Sometimes the conversation turned to issues that people of color have within the technology education profession. At one point the idea for this yearbook hit me like an out-of-control locomotive. I borrowed LaVerne's laptop and drafted an outline of a chapter for this book in the middle of one of the work sessions. The group cooperatively provided feedback, which resulted in a formal proposal to the CTTE Yearbook Committee. Dr. Everett N. Israel, who chairs the Yearbook Committee, has been extremely supportive of the book and of my editorship. For that I am very grateful.

This book would not be a book, of course, without the authors. Their commitment to their writing and diversity in a profession where almost no diversity exists has made this book a starting point for ongoing dialogue regarding diversity. I have enjoyed working with the authors and appreciate their contribution to not only the book, but the CD-ROM as well. Thank you. This book would also not be complete without the services of Dr. Lynnette Porter, the quickest and most outstanding technical editor one could ever hope to work with. I am not one to worry about minor details and am eternally grateful to those people like Lynnette who seem to find it fascinating.

I am grateful for the help and encouragement throughout this project of my husband, Dr. W. Mike Sherman, and daughter, Lindsay Sherman. I could not ask for a more supportive family. The late William E. and Anne D. Rider were both fine educators and wonderful parents; their influence was incredible. There are also many women in my family in the generations who preceded me that were strong, vital individuals who overcame adversities with dignity, humor, and love. Thank you for being such great role models and mentors. I also owe a sincere "thank you" to my friends, who have helped me to see this project through; I promise not to volunteer for this again.

The late Woody Hayes, former football coach at The Ohio State University, was attributed with the saying, "You can never pay anyone back for the help they have given you; you can only pay forward." Because I cannot pay back the technology education profession for the many wonderful experiences I've had and friends I have gained, I can only pay forward. This book is a down payment for future technology educators and, I hope, a more diverse group of professionals.

Society, Diversity, and Technology Education

INTRODUCTION

Donna K. Trautman
Bowling Green State University

Society, diversity, and technology education. Although these words may seem to indicate separate interests and issues, a thread is common to all—change. Change is inevitable, not only in each of these areas but in every individual's life. As society changes and our population in turn becomes more diverse, technology educators must adapt to the changing world. Martin (1995, p. 6) told potential technology educators that "(c)hange has eternal life and . . . those professions that do not see a need to change or simply do not adapt fast enough to change . . . become critically wounded." We must examine the society, diversity, and our profession's own mission to change. In this chapter, I examine some important demographics of our global society, the effects diversity will have on the social and economic support systems, and the status of technology education in comparison to society and diversity.

WORLD DEMOGRAPHICS

We live in a global society. The technological effects of the world on individual societies is perhaps most evidenced by the impact of technology and technological devices for work and personal use, the changes in governmental and economic support systems, and the global mobility of people, goods, and materials. To understand some world influences on the technological world infrastructure, we should view some key demographic trends and figures that will provide insight to diversity and technology education issues.

According to the Population Reference Bureau, in 1994 there were an estimated 5.6 billion individuals on earth (Population Reference Bureau, 1994). The world's population will reach six billion in 1998, and seven billion by 2008 (Population Reference Bureau, 1992). The distribution of the population has changed very little over time, as most of the world's population has lived in Asia (Population Reference Bureau, 1994). The United Nations (1992) has predicted that this distribution will continue: By 2000, the U.S./Canada will be home to 4.7% of the world's population; the Europe/Former U.S.S.R. will have 13%; Asia, 59.2%; Latin America, 8.6%; and Africa, 26.2%. Only .5% of the world's population will live in the Oceania area, including Australia and New Zealand (United Nations, 1992). With an increasing population on a limited land mass, the global society should become increasingly important. The need to share technology, make decisions about technology, and educate people to work with each other and technology will become increasingly more important for success in business, industry, and government, as well as to our very survival.

Of the 5.6 billion world citizens in 1994, 2.8 billion were women, with approximately 1.4 billion in the 15- to 49-year-old range. It is estimated that over the next 20 years this group will increase by 30% (Carty & Cryer, 1995). In 1990 an estimated 905 million adults were illiterate, of which nearly 66% were women. In a global society, more people, but especially women, will need to increase their literacy. Based on the female population, the expanding roles of women, and the demands on the global workforce, literacy of women must become a major concern. Literacy for all people is very important, and its effect on the world economy is greatly underrated: "(M)any educators, employers, training managers, and public-policy makers are not aware of the complex links between the structure of an economy and the literacy it requires" (Passmore & Mohamed, 1995, p. 8).

In part because of societal mores and low literacy rates in some countries, fewer women than men are employed. Approximately one half of the world's population of women aged 15 and older are employed in paying jobs, compared to 90% of men aged 25 to 54. These figures do not include women working domestically (Carty &

Cryer, 1995). It is imperative that women are increasingly prepared for not only the global workforce but for the consumer and ethical decision-making demands of this society. Technological literacy—for all young learners as well as all adult learners—is essential to help prepare women to meet these global needs. In *Technology for All Americans, A Rationale and Structure for the Study of Technology,* technological literacy is defined by three major areas: the ability to use technology, the ability to manage technology, and an understanding of technology (International Technology Education Association (ITEA), 1996).

The global society is also affected by the migration of people from rural areas to urban areas. In the early 1900s, only 10% of the world's population lived in urban areas, whereas in 1994, 43% of the world's population lived in urban areas (Population Reference Bureau, 1994). Urban areas are expected to grow, affecting all systems in society, including education, government, transportation, and the economy. Literacy, employment, and migration affect the quality of life. The quality of an individual's life is affected by his or her ability to participate in the global workforce, the ability to make decisions regarding career direction, the choice of environment and living location, and the satisfaction of being a productive member of society.

Developing and developed countries both affect the environment, resource consumption, rate of population growth, world labor supply, and the quality of education (Bloom & Brender, 1993). Developing countries have a profound effect on developed countries because of a high birth rate, unattended environmental issues, and little or lack of participation in the world economy. For continued global enhancement, developed countries should place a high priority on assisting developing countries to form stronger systems, including an education system. An important part of this educational system should be technology education. The opportunities for technology education to contribute to the progress of people in developing countries could be enormous. The global society would grow positively and prosper because all individuals would become educated about technological systems.

Society, Diversity, and Technology Education

A global economy has developed over the last 50 years based on increased international trade, relocation of jobs, and capital mobility. Capital mobility refers to the technological ability for capital to be transfered from one country to another almost instantaneously. This global economy is thriving because of these technological and institutional changes (Bloom & Brender, 1993). Countries are sharing knowledge and are learning from each other, businesses are able to expand across borders, and the communication systems are advancing to support these endeavors. However, at this time, because of the lack of assistance to educate all of the people about technology and the global society, the least developed countries have yet to reap the benefits. This information also supports the need for technology education to expand its influence globally with the support of a diverse group of technology educators.

Change in the global population affects the economy, workforce, society, education, and thus world citizens' future. The continuous change in technology and the population calls for educators to respond. Technology educators must respond not only to the needs of individuals, but also to global needs. Building a diverse group of technology educators, who are savvy to these diverse needs, will assist us in making the technology education curricula respond to changing global demands.

Diversity

The world's population is diverse. We typically think of diversity in terms of gender and race, but it also includes physical limitations, religion, sexual orientation, and inter-country regional cultures. As global citizens, we cannot escape diversity, nor should we want to. Although all types of diversity are important and should be considered in terms of improving technology education, in this chapter I discuss only race and gender.

In the United States, the population is becoming more diverse daily. Metaphors, including "melting pot," "salad bowl," and "mosaic," are regularly used to describe this diversity. Racial diversity is the most obvious change in U.S. society. According to the 1992 Population Bulletin, *America's Minorities—The Demographics of*

Diversity, in 1900 there were five self-reported races listed in the United States Census. In the 1990 census, there were 19 self-reported races. Minorities defined in this census included Blacks, American Indians, Eskimos, Aleuts, Chinese, Filipinos, Hawaiians, Koreans, Vietnamese, Japanese, Asian Indians, Samoans, Guamanians, Other Asians and Pacific Islanders, Other races, Mexicans, Puerto Ricans Cubans, and Other Spanish/Hispanics. Immigration is the main source of the changes in the population (O'Hare, 1992).

Between 1980 and 1992, the minority population in the United States increased by 40%, whereas the Non-Hispanic or White population grew by only 6%. This trend reflects our society's increasing need to enhance the educational levels of minorities, which now is imperative for the continued success of the United States' economy and thus the global economy. Compared to Non-Hispanic Whites, a smaller percentage of minority students graduate from high school; even smaller percentages of minorities earn college or postgraduate degrees. In the workplace, minorities receive less financial compensation for the same work as their White counterparts. Whites and Asians are more likely to hold white collar jobs and are less likely to work in the lower-paying jobs, such as semi-skilled or service jobs (O'Hare, 1992). These figures are not only disappointing but pose a threat to our society's welfare; therefore, the continued upgrading of skills and education for minorities should be a major educational concern (O'Hare, 1992).

Gender is another area of inequality in the global society. Women head one third of the world's households; however, women hold only 12% of managerial or administrative jobs. In general, women hold as little as one third of all paid jobs, and their wages are lower than men's for comparable work. In the United States between 1987-1992, women earned only 78% of the wages men earned in comparable nonagricultural employment (Carty & Cryer, 1995). Based on the information that the number of working women is increasing, women are a major consumer base, and women's proportional responsibilities of household income are increasing, the demand for educators, specifically technology educators, to respond cannot be ignored.

Gender and racial inequality in employment can often be traced to inequality in education (Carty & Cryer, 1995; O'Hare, 1992). Because the global society is increasingly technological, education for all students is correspondingly crucial. Technology education, as a profession, lags behind other disciplines in encouraging women and minorities to enter the field of technology education or technology-related coursework or professions.

The demographic data cannot be ignored, as they profoundly affect social, political, economic, and educational systems. Change, the common thread among society, diversity, and technology education, is evident when we examine the diverse population in the United States.

Technology Education

Technology education has struggled to change its image from industrial arts to its current title; however, some literature still describes programs that provide only industrial arts experiences. However, today all technology educators must respond to the needs of the global society. The Technology for All Americans Project (1996) perhaps has the most current and encompassing rationale and structure written for technology, technological literacy, and technology education. It is based on many and various inclusive perspectives regarding technology, and perhaps most importantly, it was conceptualized and written with diverse learners in mind. The Project identifies the goal of technology education as technological literacy. This document challenges technology educators in many ways, but more importantly, it calls for technology education to become an essential subject for all U.S. students. This call for action has a far-reaching potential around the world to help all individuals to become technologically literate.

The ITEA also took the initiative and developed a mission statement in its strategic plan (1993):

> ITEA promotes excellence in technology teaching and works to increase the effectiveness of educators to empower *all* people to understand, apply, and assess technology. (emphasis added)

The ITEA's strategic plan also listed four goals, one of which is the organization's resolve to "enhance participation of minorities and women in technology" (p. 6). Both the Technology for All Americans document and the ITEA mission statement provide direction for current educators and hope for future educators. However, these documents must be acted upon not only by the profession as a whole, but by individuals in leadership positions, or the documents become only words without meaning.

"This field plays a major role in preparing the future technical workforce in the United States, a workforce that is becoming increasingly diverse" (Erekson & Trautman, 1995, p. 40). Projections regarding the changing demographics of the world population and its growing diversity have been published for years; however, the changes in personnel, curriculum, and delivery to increase diversity in technology education have not been significant. The personnel developing and delivering technology education look much the same as the they did 10 to 20 years ago, even though students look much different. In 1988, Erekson and Gloeckner found that 90.38% of the technology faculty were White, 5.84% were African-American, 1.9% were Asian, .6% were Hispanic, and .5% were Native American. Approximately 1.65% of the technology faculty at predominantly White universities were African-American, and the other 4.19% were teaching at historically Black colleges. Unfortunately, a decade later, these figures are not significantly different.

The other chapters in this yearbook illustrate many issues about diversity. The authors with chapters in Section 1 investigated the history of women and African-Americans in technology education and provided some insight into a different perspective of the history of our field. Section 2 includes chapters on women technology educators and minority technology students and addresses the issues of these groups. The chapters in Section 3 provide direction for us as educators. The important topics of curriculum issues with respect to diversity, mentorship, effective leadership, and environmental and climate challenges provide us with positive initiatives to succeed in the challenge of change.

SUMMARY

The future of technology education and its educators relies on our commitment to change. The goal for this yearbook is to help provide the motivation and direction for change, and readers are encouraged to make a personal commitment to change as it relates to increasing the diversity of technology educators and effectively delivering the subject of technology to diverse students of all ages. Change means that we improve educational environments to allow all students and teachers to thrive and learn about technology. Change means that we include historical accounts of technology education that appreciate and describe the contributions of minorities and women. Technology education should be a discipline that thrives on change, but we seem to lose perspective when it comes to dealing with changes in our constituency or welcoming individuals into our ranks who are different from us. Now is truly a time for change.

REFERENCES

Bloom, D. E., & Brender, A. (1993). *Labor and the emerging world economy.* Population Bulletin, *48(2).* Washington, DC: Population Reference Bureau.

Carty, W. P., & Cryer, S. (1995). *PRB media guide to women's issues 1995.* Washington, DC: Population Reference Bureau.

Erekson, T. L., & Gloeckner, G. W. (1988). *A descriptive analysis of factors related to university employment in industrial education.* NAITTE Professional Monograph No. 3.

Erekson, T. L., & Trautman, D. K. (1995). Cultural diversity and the professions in technology. *The Journal of Technology Studies,* XXI (2), 36-42.

International Technology Education Association. (1993, April). ITEA strategic plan. *The Technology Teacher,* 52(7), 5-7.

International Technology Education Association. (1996). *Technology for all Americans: A rationale and structure for the study of technology.* Reston, VA: Author.

Martin, G.E. (1995, May/June). Professionalism: More than a mere commitment. *The Technology Teacher,* 54(8), 3, 5-6, 46.

Murphy, E. M., & Lang, A. (1994). *World population: Toward the next century.* Washington, DC: Population Reference Bureau.

O'Hare, W. P. (1992). *America's minorities—The demographics of diversity.* Population Bulletin, 7(4). Washington, DC: Population Reference Bureau.

Passmore, D. L., & Mohamed, D. (1995). Labor force literacy and the economy. *The Journal of Technology Studies,* XXI(2), 8-14.

Section I: HISTORICAL INFLUENCES OF UNDERREPRESENTED GROUPS IN TECHNOLOGY EDUCATION

A Historical View of Women's Roles in Technology Education

Karen F. Zuga
The Ohio State University

As in the greater society, women's roles in technology education are constantly changing. However, in neither case is the change a straight linear progression toward equality. Interestingly, the history of the field reveals more involvement of women in industrial education and the industrial arts of the first decades of the twentieth century than was the case during mid-twentieth century and, perhaps, today.

When I decided to seek industrial arts teaching certification in the 1970s, women in the field were an anomaly. I thought that I was breaking new ground during an era of great concern for including "nontraditional" people in all professions. I was quite surprised with my then 60-year-old father's response to my announcement. "Oh," he told me, "my manual training teacher was a woman."

For years I searched for evidence of women in the field and found few references to their existence and contributions in the mainstream histories of the field, although Bennett (1937) did write about Emily Huntington and Grace Dodge, and the often-quoted "Bonser" definition of industrial arts appeared in a book written by Frederick Gordon Bonser and Lois Coffey Mossman (1928). This led me to believe that there were women who had been involved in the shaping of technology education. In this chapter, I examine the societal roles of women and the contributions of women throughout the industrial education, industrial arts, and technology education eras in the history of our field.

SOCIETAL EXPECTATIONS OF WOMEN

While manual training and industrial education were being introduced into schools, a feminist movement was growing in the United

States. Disavowing the traditional "back seat" of second class citizenship, a few women, led by Elizabeth Cady Stanton and Susan B. Anthony, united to press for women's right to vote in every state, just as in Wyoming, which had enfranchised women in 1869. Universal suffrage was finally granted in 1920, about the time the profession adopted the name *industrial arts*, yet women were quick to realize that getting the vote did not mean gaining equality, as they still suffered a loss of property rights through marriage and were generally discriminated against in many ways. Today, women in the United States have yet to see their country endorse an equal rights amendment.

Roles During the Industrial Education Era

Women around the Western world began to assert their ideas about the right to vote and to be full and equal citizens of their nations in the second half of the nineteenth century. Historically placed in second class citizenship with no voice in government and no rights within government, women were at the mercy of their fathers, brothers, uncles, and husbands, who maintained a tight control by denying them the right to determine their own economic and political futures. Women's roles were to serve men as obedient wives, mothers, and sisters. Women who remained single were placed in lifelong guardianships in order to have a male protector who would control their finances and make important decisions for them, and widows who did not remarry often had their inherited property revert to the control of their closest male relative (Miles, 1990).

Unable to take part in the business decisions of the family or country, a handful of women began to establish themselves in nontraditional occupations and professions and to develop occupations and professions for women. From a history of child care as mothers, nannies, and dame school teachers, women easily entered professional education at the elementary school level (Danylewycz, Light, & Prentice, 1990). Related to the role of taking care of others, women carved out niches for themselves in the economy as nurses and secretaries, creating "female" professions that always reflected the status of their practitioners through their paychecks. More rare were the few women who managed to brave a hostile system of higher education in order to become doctors, lawyers, chemists, and other kinds of professionals, often with the result of being barred from practicing their chosen profession or moved into "acceptable" women's roles such as the teaching of domestic science to other women (Bellamy & Guppy, 1990).

One of the more misunderstood aspects of women at work is the role that women have always played in manufacturing. Those jobs are often thought to have been reserved for men, but many women worked in factories on assembly lines using all manner of tools. Just as some contemporary business executives are searching for lower labor costs by exporting their work abroad, women of this era were often hired in many kinds of manufacturing jobs as cheaper labor than men provided (Miles, 1990). There is a historical record, through text and photographs, of women at work in industry throughout the industrial education time period. And, just as we know that women participated in the war effort during World War II, women answered the call to fill men's jobs in factories during World War I.

At the end of the nineteenth century, the feminist movement in the United States was well underway, having inherited an organizational structure from the abolitionist movement of the Civil War era (Miles, 1990). As women in other countries began to get the right to vote, women in the United States intensified their efforts with respect to both suffrage and equal rights. They met serious opposition to their requests, as both men and other women lined up to support the status quo. As a result, the number of women who entered nontraditional professions remained few, and most women who entered teaching taught in elementary schools or were relegated to teaching domestic science in secondary schools and colleges.

Roles During the Industrial Arts Era

Just as industrial arts was gaining a hold as a subject matter in schools, women gained the right to vote in the United States in 1920, years after women in Australia, Denmark, Finland, Iceland, Norway, the Soviet Union, and Great Britain had been enfranchised (Miles, 1990). However, the right to vote did not include equal rights, and women remained in second class citizenship as their rights remained abrogated. Jobs in professions such as higher education, law, and medicine remained elusive to women, as society generally accepted the belief that women belonged in the home, taking care of family. Actually, poor women often worked out of necessity, either in factories or at home; only middle class and wealthy women could afford to stay home to care for their family needs.

Women gained a modicum of freedom during the period immediately following World War I. With the right to vote came some superficial freedoms for women, such as a growing public acceptance of more diverse roles for them. While this was happening, a growing conservative trend that would work against the new and hard won rights was in the making. With the Great Depression, society in the United States quickly reverted to a conservative approach to women's roles in the workforce, as scarce jobs were allocated to men (Faludi, 1991).

In fact, one theory is that gaining the right to vote actually worked against women, as a conservative backlash slowly reversed women's participation in the economy, education, and government (Faludi, 1991). Periods of women's social, economic, and legal advancement can be correlated with corresponding periods of conservatism with respect to the behavior and rights of women. In the twentieth century these periods have followed women gaining the right to vote, taking a major role in war production during World War II, and promoting open and widespread feminism in the 1960s and 1970s.

Throughout the industrial arts era, these periods of openness for women and backlash ebbed and flowed. The early part of the century opened with women gaining political power and beginning to take their place in the economy through a one step forward, two steps backward path and ended with women reasserting their economic rights. As fitful as the progress was, there were continual gains as women began to enter the workforce in greater numbers, both in the professions and blue collar work; to gain economic independence from male control; and to have fewer social restrictions on their personal freedoms. As each generation rid itself of a traditional restriction such as the loss of property rights in marriage or gained a fragile toehold at work by entering the professions of law, medicine, and higher education, women's limited participation in the economy increased.

At the end of the industrial arts era in the 1960s and 1970s, women were strong political activists involved in anti-war protests, promoting a wide array of women's rights, and crusading for an equal rights amendment to the United States Constitution. Moreover, traditional work roles were breaking down as more and more women entered nontraditional occupations. Women began to break down barriers in previously exclusive male occupations, such as police work, fire fighting, and the construction trades. After years of women's limited participation in the field, a handful of women in the United States were beginning to teach industrial arts in all grades.

Roles During the Technology Education Era

As the protests of the '60s and early '70s' began to die down, the hopes for an equal rights amendment began to fade, and conservatism once again dominated life in the United States. Just as the media drove women out of the factories to make way for returning veterans after World War II, women were once again subtly encouraged to return to hearth and home by such myths as the "biological clock" and the "nesting" urge. Clothing designers once again returned to exaggerated feminine images in their clothing and make up, and Hollywood created images of good mothers and homemakers. Conservative writers and the media began to extol the virtues of tending to family, as the very women who urged others to return to their nests worked (Faludi, 1991).

What those who were unwittingly engineering the conservative mood did not count on, however, was that the economy would no longer be kind to families with a single wage. Two adult salaries in a family were rapidly becoming a necessity for the middle class as well as the poor. In addition, divorce rates were higher, and women often had to rely on themselves for the family finances. Despite the images in the media, women were in the workforce to stay.

Women also made headway in all professions, not just nontraditional occupations. Traditional female professions and occupations were lower paying, and many women sought to move beyond the "pink ghetto." Unfortunately, as great numbers of women moved into some professions and occupations, such as social work, they had the negative effect of feminizing the wage scale, which resulted in lowering it. The fight to gain equal pay had still to be won as national data continued to show women at a distinct economic disadvantage.

Taking up the contemporary concerns of equal rights and equal pay, feminists focused on issues related to diversity. Although women had made progress in this century toward the goals of equality in treatment, they still had not attained their goals of an equal rights amendment, equal pay for equal work, and a lack of discrimination. Based upon these concerns and experiences, contemporary feminists have tried to promote an acceptance of diversity in knowledge, in action, and of people in order to create an environment for achieving their goals (Flax, 1990).

Despite the efforts of some women to reach goals of equality and an acceptance of diversity, many other women have not been concerned or expressed concern about these issues (Faludi, 1991). Independent of their attitude, the majority of women and families have benefited from these struggles because of a slight increase in women's wages, the right to retain property after marriage, and the right to work at a profession of choice. Today, a traditional family with both partners working benefits from a woman earning a respectable wage for her work. Staying home to care for the family may no longer be possible for most women as their roles continue to evolve. The press of the economy continues to put new demands on women and their partners to evolve.

WOMEN'S CONTRIBUTIONS TO THE EVOLUTION OF TECHNOLOGY EDUCATION

Women were more prevalent in the profession in the early years of technology education than they are today. The evidence indicates that women were fundamental to the adoption and spread of industrial education in elementary schools, and they contributed to the creation of industrial arts. However, women have not been clearly identified in the historical record of the field, and the record of their participation in the mid-twentieth century is difficult to trace.

Women's Contributions to Industrial Education

Preludes to the earliest industrial arts programs were based on a belief in the value of an "industrial" education for underprivileged children (Bennett, 1937), a growing recognition and practice of manual training in private secondary schools for liberal educational purposes (Woodward, 1898), and Froebel-inspired kindergarten programs (Herschbach, 1992), all of which gained in popularity during the nineteenth century. Elementary school industrial education and manual training programs differed from the secondary school manual training emphasis on wood, drawing, and metalworking classes for boys and directly preceded the beginning of industrial arts. These classes were labeled as industrial education. Industrial education programs for elementary schools originated in kindergarten classrooms; were aimed at instruction designed for all children, both boys and girls of all socio-economic classes; and incorporated a broad range of laboratory-based activities such as block building, drawing, book making, embroidery, crocheting, paper folding, and construction (Bennett, 1937). A number of different approaches to the teaching of manual training at the elementary school level, inspired by Swedish and British educational programs, such as arts and crafts and handicrafts, soon emerged. These educational programs banded under the name *industrial education*, where the term *industrial* was not an indication of the trades or a study of industry per se, but equated with being industrious or occupied (Bennett, 1937).

Unique to these industrial education programs were the participation and influence of women such as Emily Huntington, an industrial school administrator, and Grace Dodge, an educational reformer. According to Bennett (1937), Huntington and Dodge teamed up to create a forerunner of the Industrial Education Association, the Kitchen Garden Association, in 1880. This association was created to promote industrial education in elementary schools. It was later disbanded and became part of the newly founded Industrial Education Association in 1884 (Bennett, 1937).

Grace Dodge was an active philanthropist and educational reformer. She encouraged the practice of industrial education by promoting the ideas of Emily Huntington and securing funding for industrial education programs. Of Dodge's role in the Industrial Education Association, Bennett (1937) said,

> This new organization brought men as well as more women into the work. General Alexander S. Webb, president of the College of the City of New York, was elected president, and Grace Dodge, vice-president, *though she did the active work of a president.* (p. 413) [Italics added]

Dodge was also an active philanthropist and fund raiser. She was instrumental in raising the funds to open one of the first manual training teacher education institutions in the country, Teachers College Columbia. Through this institution much of the early professional literature of industrial education and industrial arts was created.

The early industrial education movement was influenced by a number of women who came from two educational communities: elementary school education and home economics education. Of the 188 members listed in the 1899 *Proceedings of the Eastern Manual Training Association*, 41, roughly 22%, were women. This number of women could very well have been higher, but because of the practice of listing many names with initials for all but the surname, an accurate count is not possible. Two women, Pinney and Woolman (*Proceedings of the Eastern Manual Training Association*, 1899), had papers included in the proceedings, and it is evident that both women were interested in primary school industrial education as manual training. Woolman focused on domestic arts, whereas Pinney presented a view of manual training which was inclusive of both domestic arts and those aspects of manual training such as woodworking which have been more closely associated with content for boys.

Industrial education books were published with women as *single authors* by Walker (1901), Dopp (1902), and McGaw (1909). Heller (Trybom & Heller, 1908) published an industrial education book with a male co-author. Dewey's early ideas influenced elementary school industrial educators such as Dopp (1902), as she indicated in her text by acknowledging him. Women's views about the purpose and role of industrial education differed from the mainstream male views in that women addressed elementary school education as education for *all* students.

Women also became teacher educators and department chairs at normal schools and colleges. Women such as Alice Boardman, who was hired as a supervisor of manual training at the Michigan Normal School in 1901, eventually led some of the manual training and industrial arts departments. Boardman was instrumental in creating the Department of Industrial Arts at what eventually was to become Eastern Michigan University (*College of Technology Annual Review, 1994-1995*).

Women's Contributions to Industrial Arts

Many female elementary school educators actively contributed to the role of manual training, industrial education, and, eventually, industrial arts in the elementary school curriculum. These women who were involved in the industrial arts movement were both teachers and teacher educators, and they contributed to and authored articles in conference proceedings and journals as well as wrote textbooks on the subject of industrial arts education for children. Contributions to the *Teachers College Record* were made by many women, among them Weiser (1907), Hennes (1921), Bennett (1911), and Watkins (1911). Curiously, in several articles that appeared in *Teachers College Record*, authorship has been identified as "the Teachers of the Horace Mann Elementary School" (Russell,

1913), making identification of women's writing more difficult. Nonetheless, a quick survey of a single journal during the formative years of industrial arts education reveals that women actively contributed to the professional literature. Another difficulty in tracking women's work and writing was the tendency for many teachers to produce curriculum units as published or unpublished work. Newkirk (1940) identified many examples of this kind of publishing on the part of women in his book about handwork for elementary schools.

Women who wrote about the goals and purpose of elementary school versions of industrial arts advocated a social reconstruction agenda for industrial arts. Wiecking (1928) stated

> Intelligently planned and directed, the field of manual training and industrial education may change attitudes and foster appreciations of great moment in the child's life. With a background of actual participation in manual activity there may come to the child a point of view about our civilization which is bound to affect his [sic] future behavior toward questions arising in industrial life. (p. 268)

Wiecking echoed the sentiment of the time and of those who practiced elementary school forms of industrial arts. There was always a greater purpose than prevocationalism and skill training. The skills of concern were general life skills, rather than particular vocational skills. Hennes (1921) wrote

> There are certain social ideals and skills absolutely necessary in order to live unselfishly and helpfully in society with their fellows. Our children must learn how to cooperate. They must learn the spirit of mutual helpfulness. They must come to appreciate fair play, and thus become unselfish in their dealings with others. They must, in particular, learn to be truthful. (p. 137)

Men also contributed to elementary school industrial arts literature. Bonser collaborated with Mossman (1921) and wrote generally about industrial arts, incorporating information for elementary school practice as well as junior and senior high schools (Bonser,

1930). His prescriptions for content, based upon his original work with Mossman, differed from contemporary texts written solely by men. McMurry, Eggers, and McMurry (1923) published *Teaching of Industrial Arts in the Elementary School* and focused only on woodworking and bookbinding as the content for elementary school industrial arts. Most of their text was devoted to the traditional design and planning of wood projects. Winslow (1923) provided a more general view of curriculum content with bookmaking, paper making, the manufacture of baskets and boxes, brick and tile, pottery, cement and concrete, textiles, copper, iron and steel, soap, glass, and wood as topics of study, yet he did not approach content comprehensively by dealing with technologies of the home, in particular, the growing and preparation of food. His focus was on industry and the industries represented by his topics. Comparisons of industrial arts texts written by men and those written solely by women or co-authored by women and men reveal differences in the way in which industrial arts content was approached by women and men. Those texts written by or influenced by women during the early part of the century tend to have a more general scope with respect to content, incorporating all technologies, those of both industry and the home.

Influenced by Dewey, perhaps by association at Teachers College, Bonser and Mossman produced one of the first industrial arts texts for teachers, *Industrial Arts for Elementary Schools*. In that text they provided the emerging field of industrial arts with the definition of the subject matter that would carry them through the first half of the twentieth century. Of the industrial arts, they said

> The industrial arts are those occupations by which changes are made in the forms of materials to increase their values for human usage. As a subject for educative purposes, industrial arts is a study of the changes made by man (sic) in the forms of materials to increase their values and of the problems of life related to these changes. (1928, p. 5)

By the time that this book first appeared in the 1920s, the definition of the term *industrial education* had become associated with vocational education and was explained by Bonser and Mossman

(1928) as "a definitive, intensive training for productive work in some industry" (p. 6). Industrial arts was distinguished from industrial education (as vocational education) as a study with general education purpose. Materials, tools, processes, and production were to be studied "for the values which such study affords in one's everyday life, regardless of his [sic] occupation" (p. 6).

More important was the justification and purpose for the study of industrial arts, which incorporated the ideas offered by Dewey in his discussions of the study of the occupations (Dewey, 1916) and social reconstruction (Dewey & Childs, 1933). The purpose of industrial arts was, according to Bonser and Mossman, to study "such problems of citizenship as to share in the regulation of industry" (1928, p. 7), which relates directly to Dewey's (1916) ideas about the social role of occupations in the curriculum, of which he stated, "The most direct road for elementary students into civics and economics is found in the consideration of the place and office of industrial occupations in social life" (p. 201).

Dewey did not work in isolation. He had been influenced by a woman of some note, Jane Addams. As a pioneering social worker, Addams influenced not only Dewey, but the era in which industrial arts had become a subject in the nation's schools. In *Democracy and Social Ethics*, Addams addressed the role of education in providing a social value to children's school experiences for the purpose of reconstructing American life (Addams, 1902).

As a result of the established practices of industrial arts education in the elementary schools and the influence of Dewey upon educators in general, as well as leaders such as Addams, during the first half of the century industrial arts was given a place in the schools (Zais, 1976). Often, that place was to be a vehicle for the study of the occupations that permitted integration, acquisition, and application of practical knowledge to social problems. University laboratory schools such as The Ohio State University Elementary School and Kindergarten and public schools began to establish laboratories to provide "for real participation by each student in each of these functions of living" (Publications Committee, 1935, p. 121).

Although social reconstruction was the direction and intention of many elementary school and industrial arts educators of the time, their influence was not the mainstream direction taken by the industrial arts community as it moved forward into the twentieth century. Most voices advocating social reconstruction in the industrial arts curriculum gradually became silent as the industrial arts curriculum theorists focused on identifying better ways to teach skills (Selvidge, 1923; Selvidge & Fryklund, 1946; Fryklund, 1956) and unique content for industrial arts (Towers, Lux, & Ray, 1966). School practice became more vocational with a curriculum of woodworking, metalworking, and drawing (Schmidtt, 1963).

As the era of the industrial arts progressed, women's voices fell silent. Evidence of participation in the profession by women grew scarcer, and they disappeared from the ranks of industrial arts teachers. As industrial arts evolved, it became apparent that early interest in providing a subject matter for all children, both boys and girls, was not being practiced as the common pattern of industrial arts for boys and home economics for girls became set. In the literature for the field, prescriptive theory for content became more narrowly focused on woodworking, metalworking, and drawing in secondary schools (Schmidtt, 1963), and the inclusive curriculum of elementary schools focused on weaving, sewing, clay modeling, woodworking, block building, and paper construction (Wiecking, 1928). The emphases on foods, clothing, shelter, utensils, records, and tools and machines (Bonser & Mossman, 1923) were diminished. Some reasons for this evolution are the association with vocational education in collegiate and secondary education and the loss of the female elementary educators' voices.

During mid-century a few women did continue to participate in the American Industrial Arts Association (AIAA) as elementary education representatives. However, their voices disappeared from the mainstream literature of the industrial arts. Correspondingly, industrial arts almost disappeared from elementary schools.

Women's Contributions to Technology Education

As technology education evolved, women began to reappear as leaders and in the professional literature. Before industrial arts professionals were ready to accept the redefinition of their field of study as technology education, several women who were working in elementary school education took up the suggestions of Warner et al. (1965), Olson (1963), and DeVore (1964) and began to implement technology education programs in the schools. Elizabeth Hunt, Mary Margaret Scobey, and Norma Heasley all instituted technology education projects that preceded the implementation of technology education by the mainstream profession.

Just as Lois Mossman Coffey with Frederick Gordon Bonser had introduced a groundbreaking book about industrial arts in the 1920s, Mary Margaret Scobey published one of the first technology education books in 1968, *Teaching Children About Technology*. In her book, she also incorporated a contemporary curriculum structure that included construction, cybernation, communication, food processes, textiles and clothing, power, and transportation. The book was written to prepare elementary school teachers to teach about technology in the elementary school. Scobey also contributed a chapter about industrial arts for the elementary school to the American Council on Industrial Arts Teacher Education (ACIATE) Yearbook (1974).

Working in the elementary schools in New Jersey and aided by career education funding during the 1970s, Elizabeth Hunt introduced a technology education program called Technology for Children (T4C). This program focused on technology education activities as well as career education (Cochran, 1970). It resulted in a lasting commitment in the New Jersey elementary schools to technology education. It also became a widely studied program, and several students wrote dissertations that analyzed various aspects of the program.

Following Hunt's lead and inspired by industrial arts classes as an education student, Heasley, in conjunction with her husband, an industrial arts teacher educator, created a program in Northeastern Ohio entitled A Technological Exploratorium (Cochran, 1970). It was also funded by federal and state career education moneys, which were readily available during the 1970s. Heasley (1974) also wrote about designing industrial arts curriculum for elementary schools in the 23rd yearbook of the ACIATE.

These women, as well as others, were early advocates of technology education. Just as the women who had preceded them in contributing to the definition and implementation of industrial arts, they were associated with elementary schools and elementary school teacher education. As women's roles were beginning to change in society, these women also received greater recognition of their contributions to the profession by the members. Scobey was a founding member and the first president of the American Council for Elementary School Industrial Arts (ACESIA), which became incorporated as a part of the AIAA. (ACESIA is now called the Technology Education for Children Council and remains as a council of the International Technology Education Association.) Heasley also served as president of ACESIA. Both women served on the Board of Directors of the International Technology Education Association as representatives of ACESIA. As a result of her involvement in the profession, Heasley was included as a member of the Jackson's curriculum team in 1980 Mill (Snyder & Hales, 1981).

As these women were breaking ground in technology education, a number of younger women were preparing to become industrial arts teachers. Slowly, as nontraditional roles for women were beginning to open in society, a few women began to teach industrial arts, not as elementary school teachers, but as middle school and high school industrial arts teachers. That generation of women are now assuming leadership roles within all areas of the profession as teachers, administrators, teacher educators, association officers, and executive directors.

The male traditions of the profession are changing as women begin to break ground in previously unknown territory. Nevertheless, at the university level, women were still not well represented at the department or program leadership level. Therefore, in 1990 an organization designed for department chairs and state department representatives, the Mississippi Valley Industrial Teacher Education Conference, opened five memberships for women and minorities. Today, women are breaking into the ranks of department and program leadership in technology education, and their leadership is being actively sought at universities. The effort to recruit women into the profession as teachers, teacher educators, and administrators continues to grow.

Exploring the Fragmented Record

From the accepted, traditional role as elementary school educators to more varied roles today, women have always contributed to the evolution of technology education. More than just implementing a program, they actively influenced the direction of industrial education, industrial arts, and technology education through practice, literature, and leadership. Although evidence of their participation has been difficult to locate, women have been contributing members of the profession. For the most part, early female industrial education and industrial arts teachers were elementary school teachers conforming to the roles assigned to them by society. But their roles as elementary school practitioners did not prevent them from being innovators. In fact, their roles may have helped them to view industrial arts as a subject for all students, thereby creating a broader view of the field than those who were creating a subject for the prevocational and vocational education of boys.

Women were also active in establishing professional organizations and the conduct of professional organization business. Dodge and Hunt were early advocates of professionalism as they formed the Kitchen Garden Association and later helped to found the Industrial Education Association. Scobey was instrumental in the founding of ACESIA. Today, many women carry on the tradition through service to professional associations related to technology education by serving in all offices.

Technology education professional literature has also been enhanced by the contributions of women through their writing and explanation of new types of curriculum. Mossman, Hunt, Scobey, and others created an innovative curriculum for the field long before it was adopted by the majority. It is impressive that the curriculum ideas that they advocated were eventually adopted.

As impressive as their accomplishments are, an underlying question remains to be discussed: Why is the record of women's participation so fragmented in the professional literature? Why are we just beginning to highlight the contributions of women to our field, as was done in Eastern Michigan University's *College of Technology Annual Review* of 1994-1995?

One explanation is that as industrial arts became tied to vocational education for political expediency, the historical gender split in vocational education into two camps of the trade and industry males and the home economics females left the elementary school females in a difficult position. The numbers of women had dwindled, plus the reconstituting of manual training into several vocational areas left no place for women in industrial arts.

After having been integral to the effort to initiate industrial arts, women and their voices backed out or were shut out of industrial arts and industrial education. Given the social norms of the day and the growing numbers of men who were taking control of all industrial education efforts, including industrial arts, early female advocates and practitioners in industrial arts were slowly shut out and the record of their participation in the field forgotten. Early in this century, female industrial education and manual training teachers were not anomalies in schools (Bennett, 1937), and there was evidence of their participation in the profession through the literature of the period (Newkirk, 1940).

Well-documented and recurring social pressures accentuated differences with respect to gender roles at home and at work. According to Faludi (1991), these social mechanisms countered the progress women made toward equality. Most notable was the propaganda effort at the end of World War II, initiated by government, industry, and media representatives and aimed at driving women out of industry and back to the home and hearth to make way for returning veterans. It is not impossible that a backlash as a result of women's suffrage resulted in the diminishing of women's voices in the 1930s, both in society and industrial arts.

There were also many subtle techniques for diminishing women's voices in writing. Russ (1983) in an analysis of women's contributions to nineteenth century literature observed

> The methods . . . are varied but tend to occur in certain key areas: informal prohibitions (including discouragement and the inaccessibility of materials and training), denying the authorship of the work in question (this ploy ranges from simple misattribution to psychological subtleties that make the head spin), belittlement of the work itself in various ways, isolation of the work from the tradition to which it belongs and its consequent presentation as anomalous, assertions that the work indicates the author's bad character and hence is of primarily scandalous interest or ought not to have been done at all (this did not end with the nineteenth century), and simply ignoring the works, the workers, and the whole tradition, the most commonly employed technique and the hardest to combat. (p. 5)

Several techniques mentioned by Russ can be identified in a historical analysis of the loss of women's voices in industrial arts. Often, the definition of industrial arts that was written by Bonser and Mossman (1928) has been cited in text as the "Bonser" definition (Lux, 1981; Smith, 1981), giving credence to Russ' (1983) theory of misattribution. Other factors operating during that time period included the loss of women as home economics split from industrial education and a lack of opportunity for women to study industrial arts because of the limited roles that they were permitted by the greater society.

A good example of the profession ignoring a work altogether and asserting that the work ought not to have been done at all is the work of Maria Montessori. As the industrial arts were evolving in the early part of the century, Montessori was experimenting with similar activity-based forms of education, yet her work was largely ignored by industrial arts advocates in the United States. Although there was interest in her theories and work in the United States, it was largely overlooked. Moreover, in 1914 Kilpatrick took it upon himself to critique her work, placing her concern for the individual in contrast to the progressives' concern for society (Beck, 1961). Today, having stood the test of time, Montessori's theories and methods have offered much to educators.

As women have historically occupied second class citizenship and social status, ignoring their work and their contributions to the intellectual endeavor of a field such as technology education is a relatively easy process. The record of scholarship in technology education bears witness to the veracity of this claim.

SUMMARY

Women have shaped the direction of technology education throughout its evolution as a field of study in the schools. At times, women's influence has been greater than at other times, and much of their influence has eroded over the profession's century of existence. Nevertheless, women have clearly proposed and practiced innovative forms of technology education throughout its history. Although much of the effort by women has been relegated to elementary school education as a result of socially accepted roles and opportunities for women, these roles and opportunities are changing.

Women will continue to shape the direction of technology education through leadership as teachers, teacher educators, and administrators as more and more women become technology educators. Women are once again contributing to technology education by providing a contemporary body of literature, taking an active role in association leadership, and leading educational programs in schools and colleges. Perhaps they are now here to stay as full members of the technology education profession, or perhaps they and their voices will once again be permitted to disappear from the record of the field.

REFERENCES

Addams, J. (1902). *Democracy and social ethics.* New York: Macmillan.
Beck, R. (1961). Kilpatrick's critique of Montessori's method and theory. *Studies in Philosophy and Education,* 1(4), 153-162.
Bellamy, L. A., & Guppy, N. (1990). Opportunities and obstacles for women in Canadian higher education. In J. Gaskell & A. McLaren (Eds.), *Women and education* (pp. 163-192). Calgary, AB: Detselig.
Bennett, C. A. (1937). *History of manual and industrial education 1870 to 1917.* Peoria, IL: Chas. A. Bennett.
Bennett, C. R. (1911). A study of the wheat industry in the fifth grade. *Teachers College Record,* 12(1), 50-4.
Bonser, F. G. (1930). *Industrial arts for public school administrators.* New York: Teachers College, Columbia University.
Bonser, F. G., & Mossman, L. C. (1928). *Industrial arts for elementary schools.* New York: Macmillan.
Cochran, L. H. (1970). *Innovative programs in industrial education.* Bloomington, IL: McKnight.
College of Technology Annual Review. (1994-1995). Ypsilanti, MI: Eastern Michigan University.
Danylewycz, M., Light, B., & Prentice, A. (1990). The evolution of the sexual division of labor in teaching: A nineteenth-century Ontario and Quebec case study. In J. Gaskell & A. McLaren (Eds.), *Women and education* (pp. 33-60). Calgary, AB: Detselig.
DeVore, P. W. (1964). *Technology: An intellectual discipline.* Bulletin No. 5. Washington, DC: American Industrial Arts Association.
Dewey, J. (1916). *Democracy and education.* New York: Free Press.
Dewey, J., & Childs, J. L. (1933). The social-economic situation and education. In W. H. Kilpatrick (Ed.),*The educational frontier* (pp. 32-72). New York: Appleton-Century.
Dopp, K. E. (1902). *The place of industries in elementary education.* Chicago: University of Illinois.
Faludi, S. (1991). *Backlash: The undeclared war against American women.* New York: Crown.

Flax, J. (1990). Postmodernism and gender relations in feminist theory. In L. J. Nicholson (Ed.), *Feminism/Postmodernism* (pp. 39-62). New York: Routledge.

Fryklund, V. C. (1956). *Analysis techniques for instructors*. Milwaukee, WI: Bruce.

Heasley, N. (1974). Industrial arts in the elementary school: Designing curriculum. In R. G. Thrower & R. F. Weber (Eds.), *Industrial arts for the elementary school* (pp. 88-154). Bloomington, IL: American Council on Industrial Arts Teacher Education.

Hennes, M. (1921). Project teaching in an advanced fifth grade. *Teachers College Record, 22*(2), 137-148.

Herschbach, D. R. (1992). Commencing the education of an industrial people: The early kindergarten movement. *Journal of Epsilon Pi Tau, 18*(1), 20-32.

Lux, D. G. (1981). Industrial arts redirected. In R. Barella & T. Wright (Eds.), *An interpretive history of industrial arts* (pp. 205-226). Bloomington, IL: American Council on Industrial Arts Teacher Education.

McGaw, V. (1909). *Construction work for rural and elementary schools*. Chicago: A. Flanagan.

McMurry, O. L., Eggers, G. W., & McMurry, C. A. (1923). *Teaching of industrial arts in the elementary school*. New York: Macmillan.

Miles, R. (1990). *The women's history of the world*. New York: Harper & Row.

Mossman, L. C. (1921). The project method in the industrial and household arts. *Teacher College Record, 22*(4), 322-328.

Newkirk, L. V. (1940). *Integrated handwork for elementary schools*. New York: Silver Burdett.

Olson, D. W. (1963). *Industrial arts and technology*. Englewood Cliffs, NJ: Prentice Hall.

Proceedings of the Eastern Manual Training Association. (1899). New York: Author.

Publications Committee. (1935). Certain implications of the program of the university school laboratory. *Educational Research Bulletin, 14*(5), 117-123.

Russ, J. (1983). *How to suppress women's writing.* Austin: University of Texas.

Russell, J. E. (Ed.) (1913). Contents. *Teachers College Record, 15*(2), np.

Schmidtt, M. L. (1963). Trends and developments in industrial arts. *Industrial and Vocational Education, 52*(1), 16-18, 55.

Scobey, M. M. (1968). *Teaching children about technology.* Bloomington, IL: McKnight & McKnight.

Scobey, M. M. (1974). Industrial arts in the elementary school. In R. G. Thrower & R. F. Weber (Eds.), *Industrial arts for the elementary school* (pp. 15-39). Bloomington, IL: American Council on Industrial Arts Teacher Education.

Selvidge, R. W. (1923). *How to teach a trade.* Peoria, IL: Manual Arts Press.

Selvidge, R. W., & Fryklund, V. C. (1946). *Principles of trade and industrial teaching.* Peoria, IL: Manual Arts Press.

Smith, D. F. (1981). Industrial arts founded. In R. Barella & T. Wright (Eds.), *An interpretive history of industrial arts* (pp. 165-204). Bloomington, IL: American Council on Industrial Arts Teacher Education.

Snyder, J. F., & Hales, J. A. (1981). *Jackson's Mill industrial arts curriculum theory.* Charleston: West Virginia Department of Education.

Towers, E., Lux, D. G., & Ray, W. E. (1966). *The rationale and structure of industrial arts subject matter.* Columbus: The Ohio State University.

Trybom, A. B., & Heller, R. R. (1908). *Correlated hand-work.* Detroit, MI: Speaker.

Walker, L. (1901). *Varied occupations in weaving.* New York: Macmillan.

Warner, W. E., Gary, J. E., Gerbracht, C. J., Gilbert, H. G., Lisack, J. P., Klientjes, P. L., & Phillips, K. (1965). *A curriculum to reflect technology.* Columbus, OH: Epsilon Pi Tau.

Watkins, F. V. (1911). Constructive history in the eighth grade. *Teachers College Record, 12*(1), 60-65.

Weiser, L. H. (1907). Manual training. *Teachers College Record,* 8(1), 29-34.
Wiecking, A. M. (1928). *Education through manual activities.* New York: Ginn and Company.
Winslow, L. L. (1923). *Elementary industrial arts.* New York: Macmillan.
Woodward, C. M. (1898). *Manual training in general education.* New York: Scribner & Welford.
Zais, R. S. (1976). *Curriculum: Principles and foundations.* New York: Harper & Row.

Chapter 3: Contributions of African-Americans to Technology Education

Michael L. Scott
The Ohio State University

Keith V. Johnson
East Tennessee State University

According to Hall (1973), industrial (technology) education has been a significant part of the culture of Black Americans since the time they were brought to America as slaves. Black persons of African descent were imported to America to work on cotton plantations in the South. Although Africans used technology to solve human problems in their native continent, they generally arrived in America without any specific industrial skills that would enable them to adapt to their new roles in their foreign country. From the very beginning, African slaves were being taught skills relative to the industrial and economic needs of the plantation that they served. Generally, the major goal of slave owners was to increase the worth of their plantations, and slaves were taught industrial skills to that end.

In this chapter we present an overview of the contributions of African-Americans to technology and technology education. First, there is a brief history of African-American involvement in industrial/technological education. Second, contributions of Africans and African-Americans to the origin and evolution of various technologies as they relate to technology education are highlighted. Finally, there is a brief discussion of the roles that African-Americans have and can continue to play in technology education.

Early Involvement of Africans and African-Americans in Industrial Education

To understand the involvement of African-Americans in technology education, we must start by taking a brief look at Blacks and oppression. Since Blacks were brought to what would become the United States as indentured servants in the early 1600s, there has existed a foundation for the development of industrial skills and technological competence. Moody (1980) asserted that some historians have given recognition to African-Americans who were artisans, inventors, blacksmiths, and the like. However, much of their involvement was not by choice. The technological skills that slaves learned were those that the slave owners bestowed upon them to sustain the economic potential and survival of their plantations.

The nature of what slaves learned and did was quite varied in terms of the type of plantations that they served. Hall (1973) contended that during colonial days, English settlers used an apprenticeship system to teach slaves specific skills. By encouraging slaves to acquire multiple skills, slave owners increased the worth of their property. However, a much larger number of slaves were involved in unskilled jobs such as maids, butlers, and the like (Moody, 1980). Slaves with multiple skills (and those who were highly skilled) became premium "trading bait" at slave markets (Moody, 1980).

Highly skilled slaves in some rare instance were provided opportunities to buy their freedom. Therefore, there was some motivation for slaves to become highly skilled. Additionally, these slaves taught skills to their subordinates (other slaves) and developed a sense of authority over those who worked underneath them. Still, the major purpose for learning and teaching industrial skills was for the good of the plantations and not for the good of the individual. What, then, became of these slaves after they were freed?

According to Spivey (1978), there was a "widespread belief that the newly emancipated slaves were a childlike people, inferior and unable to fend for themselves" (p. 1). Bennett (1926) made the same point by stating "at the close of the Civil War there were important Negro problems to be solved" (p. 243). The general feeling when the war was over, was that *something* had to be done about helping ex-slaves adjust as free people during the Reconstruction Era.

The Freedman's Bureau was a governmental agency charged with the responsibility to help Blacks in their adjustment (Anderson, 1982; Bennett, 1926; Spivey, 1978). General Samuel Chapman Armstrong (a White man who commanded the Eighth Regiment of Colored troops) was one of eight agents in charge of the Freedman's Bureau in Virginia. Armstrong began the (now called) Hampton Institute and became its first principal. Armstrong felt "Negro" adjustment could only occur through education (Barlow, 1967) and that skilled "Negro" labor was a must for Reconstruction in the South.

Many historians credit Armstrong with starting the Hampton Institute, but few cite the influence of his father in formulating the "Hampton" model. Armstrong's mother and father were missionaries in 1843 in Hawaii, where they instituted their Puritanic ethics on the native Hawaiians. They believed that Christian virtues of hard work and morality could be achieved through industrial education (Spivey, 1978). The Armstrongs started a boarding school called the Hilo Manual Labor School for Native Hawaiians, where the natives "paid" their expenses by doing such activities as gardening and carpentry.

Armstrong, as a boy, observed the "Hawaiian" model as a successful model of industrial education for the underclass. He also inherited his parents' attitudes regarding this underclass. His father wrote

> My general plan is to aim at the improvement of the heart, the head and the body at once. This is a lazy people and if they are ever to be made industrious the work must begin with the young. (Spivey, 1978, p. 18)

According to Spivey (1978), Samuel Armstrong (influenced by his parents) thought Hawaiians were "savage" and that Mexicans were dirty, smelly, and nasty-looking. But he felt that Blacks had a higher potential for success and therefore could be educated.

Some claim that Armstrong, in starting the Hampton Trade School, was more interested in bringing about order and stability than he was in helping Blacks advance (Spivey, 1978). Nevertheless, the Hampton Institute became an exemplar in the education of Blacks, as it still is today. One of its distinguished graduates was Booker T. Washington (who later began the Tuskegee Institute in Alabama).

According to Anderson (1982, 1988), the Hampton-Tuskegee model was promoted by Booker T. Washington as *the* model for industrial education for junior and senior high school students. In 1885 Booker T. Washington gave his famous "Atlanta Compromise" speech, which gave way to the much acclaimed Booker T. Washington and W. E. B. Dubois debates (Moody, 1980).

In this speech, Washington asserted that economic liberation, through the development of technical competence, production, and skill, would benefit African-Americans. Many viewed his stand as an attempt at "peaceful coexistence" with White people (Moody, 1980). Still, some of his ideas are expoused today: economic liberty, Black enterprise, and Black business and development.

W. E. B. Dubois, who was a Black leader at the time, felt that Washington's views would continue to repress Black people and that industrial education was a conservative bulwark to Black liberation. He felt that Blacks should be educated in the classics and that a liberal education was the way to freedom. Dubois (1932) contended that the "blending" of practical and liberal education could be empowering when united toward a cultural goal. The Washington-Dubois debates can be characterized as what Lux (1981) referred to as "learning *about*" (Dubois' view) and "learning *how to*" (Washington's view). Interestingly, Lux advocated both positions for a person to be truly technologically literate.

The second Morrill Act of 1890 provided a great source of support for some historically Black colleges and provided the impetus for supporting segregated, historically Black institutions at the same level as their White counterparts had been supported under the Morrill Act of 1862. However, Anderson (1982) noted that there were separate and unequal industrial education systems in place until the 1960s, when civil rights legislation was enacted.

Contributions of Africans and African-Americans to the Origin and Evolution of Various Technologies

Africans and Americans of African descent have contributed significantly to the evolution of many technologies. Whites have claimed credit for many technologies of African origin. Because of the insignificant value that many White Americans placed on African-Americans as individuals, many of their brilliant ideas and concepts were simply not recognized.

Further, the inventions of African slaves in America automatically belonged to their owners. It wasn't until after slavery that African-Americans were given credit for their inventions when they were patented. But even when some patents were sold to Whites, African-Americans did not receive proper credit. Therefore, it is virtually impossible to show all significant contributions of African-Americans in our society.

For example, most historians give Eli Whitney, a White New Englander who migrated south to study law and teach on a cotton plantation, total credit for the 1792 invention of the cotton gin, a machine that separates cotton fibers from its seed. Cleaning cotton by hand was extremely difficult and time consuming for slave laborers. Therefore, slaves sought to find an easier way to separate the seed from the cotton.

According to Hayden (1972), Whitney was in Georgia when he saw a comb-like instrument that separated the seeds from the cotton. It was made and used by a slave named Sam who learned this practice from his father. Whitney improved upon this concept and was given full credit for its origin. The concept revolutionized the harvest of cotton. This type of credit transition was typical of the way African-Americans were treated by Whites, who commercialized Blacks' brilliant ideas. Another example of a widely held misconception is that Thomas Jefferson built Monticello. Paulsen (1993) cited the real builders in the following:

> There's too much made of it . . . (Jefferson) didn't build Monticello . . . Isaac (a slave) did . . . he was a carpenter, he was a slave . . . and he built most of what people go there

and see. And nobody even talks about him, they talk about Jefferson . . . Jefferson couldn't hold a hammer . . . he was (technologically) incompetent . . . the point is . . . (T)he people who really built (Monticello) were Black people, were African American people. (cassette recording)

Different technologies originated in all areas of the globe by a variety of races and cultures. African-Americans' contributions to technological knowledge began in the United States or Africa. Until recently, many of these contributions were attributed to White people.

OUT OF AFRICA

Africa is recognized by the world's leading archeologists as the birthplace of the human race. The oldest known bones (perhaps some 125,000 years old) and the oldest known tools have been found in Africa. Stone Age humans first learned to make sophisticated tools about 90,000 years ago in Africa, not Europe, as many believed. That level of technology materialized in Africa about 75,000 years before it began appearing in Europe and Asia. Africans were also among the first people to create villages and cities. This technology arguably led to modern construction technology (Yellen, Brooks, Cornelissen, Mehlman, & Stewart, 1995).

In Egypt, between 3000-2000 B.C., pyramids were erected. By 2500 B.C., the Egyptian Pharaoh piloted the largest of all pyramids. The base was 756 feet square, and the pyramid rose to 482 feet. It has been calculated that two and a half million blocks of stone were required in its construction, and the average weight of each block was more than two and a half tons. Such a structure would require a considerable amount of planning, mathematics, practical geometry (perhaps descriptive geometry), and the basic elements of engineering. The Africans who were able to pull off this feat were more than just laborers. They were brilliant thinkers and technological problem solvers (Hodges, 1970).

The following examples highlight some significant technological contributions of Africans in ancient Africa. Although this is not an inclusive listing, the examples illustrate the historical development of technology in Africa. These illustrations show that beginning in approximately 2000 B.C., there were evolved civilizations in Africa whose technical developments had an influence on Europe (Hilliard, 1990).

Construction Technology

Construction technology has evolved through the advancement of modern tools and machines, making carpenters' tasks easier and less time consuming. Authors of U.S. history books generally attribute a great deal of credit for the origin and evolution of many tools, equipment, and machines to Whites, many of whom are well deserving. However, many construction technologies began in Africa, then found their way into Asia, Europe, and America.

A group of Egyptian carpenters' tools from about 1500 B.C. were recently discovered in tombs (Hodges, 1970). Because of the excellent condition of the tombs, the wood and metal objects within them survived. The Egyptian carpenters' tools had copper blades and wooden handles. Thanks to the Africans' unique artistic and graphical ability, the use of each tool can be seen in the wall paintings inside the tombs. As early as 2500 B.C., woodworking, as depicted on the walls of a tomb, had become sophisticated. Ax and adz were used for rough shaping; timber was lashed between upright supports while being sawed; mortises were cut with mallet and chisel; and final polish was given with blocks of sandstone. These tools and skills did not appear in other cultures until much later.

An early example of construction technology can be found along the Nile River in the northeast corner of Africa. There the first great kingdom of the ancient world, Egypt, was established. Menes was the first Pharoah, from about 3100 B.C. to 3038 B.C. Menes not only concerned himself with the unification of Egypt but the control of the rivers, which influenced agriculture and commerce. He is credited with the first damming of the Nile and the digging of dikes for agricultural purposes (*Black History at an Early Age*, 1991). His rule not only contributed to construction technology but also more broadly to civil engineering. These technologies have influenced the current way of manipulating various bodies of water for agricultural uses and flood control.

Manufacturing Technology

Most technology educators would agree that manufacturing technology is the study of converting raw materials into industrial standard stock, which is further processed into finished products. This process has traditionally been completed in one location. The Egyptians saw the value of manufacturing because it involved each person working independently, yet yielding a fairly complex product. An early tool used in clothing manufacture was the loom for weaving garments. Early pictures of looms have been discovered in Egyptian and Mesopotamian artifacts (exact dates unknown). Today the loom is important in the textile industry.

The invention of the lathe is attributed to Anarcharis in the third century B.C. in Egypt. Today's lathe also contributes significantly to the manufacturing industry.

Bio-Related Technology

Bio-related technology requires the use of living organisms or their parts to make or modify commercial products. Although the area is fairly new to technology education, the application of bio-related technologies has been important for a long time.

Beer-manufacturing technology existed long before modern mass production technologies evolved. According to illustrations from tombs, beer making in Egypt and Mesopotamia was done about 2500 B.C.

In addition to beer, early Egyptians made wine. Tomb paintings from 2000 B.C. illustrate early Egyptian simple presses to extract the juices from grapes or date-palms. The fruit was placed in a tall linen bag, which was then stretched between two posts. Bar handles were placed at both ends of the bag and twisted to extract the juice from the pulp. Through the biological process of fermentation, wine was created.

Ceramics Technology

Ceramics technology also has part of its roots in Africa. Ceramics technology is the study of clay techniques and applications. It can also be referred to as the art of creating new entities from the mingling and blending of clay and fire.

According to Mikami (1972), pottery was being made by people living in the Near East as early as the fifth millennium B.C. In northern Nigeria, before 4000 B.C., people discovered that clay vessels could be fired to produce more durable pottery. Manufacturing methods required no specialized pieces of equipment. The clay was mixed with sand or crushed rocks and shaped by hand over a pot. Shapes were simple and often seem to imitate skin containers or gourds, both of which may well have been used for storage and drinking vessels long before pottery was created. Like so many early inventions, we may not be able to retrace the steps by which pottery came into being; it is probable that a long period of experimentation with vessels made of unfired clay preceded the invention of pottery (Hodges, 1970).

We do know, however, that the development of ceramics took a very long time to become sophisticated. According to Mikami (1972), the first high-fired ceramics appeared in Egypt about 3000 B.C. These Egyptian ceramics were somewhat different from modern high-fired wares. The body clay was made from silica that had been ground to a fine powder, and the shaped vessels were thickly covered with copper glazes in an alkaline-flux solution. Other cultures later adopted and refined this early ceramic technology.

Transportation Technology

Transportation is defined here as any means of moving people or goods from one place to another via air, land, or water. African-Americans have contributed significantly to the conceptualization and development of many forms of transportation, including water transportation. Today's transportation via water has become sophisticated and, to an extent, a luxury available to those who can afford to participate in boating. Boating existed before 3000 B.C.; it was depicted on Egyptian artifacts and Egyptian tomb paintings about 2500 B.C.

Early Egyptian boats were built of papyrus. The bow and stern were made up of cut bundles of reed, which created a blunted, upturned end, resembling a banana. The boats were lightweight to assist in negotiating the rivers. The boats appear to have been paddled at the beginning of the third millennium B.C. Only later did the Egyptians row vessels. According to Hodges (1970), the very poor drawings from the pre-dynastic period in Egypt suggest that some early boats were sailed, via a single mast supported by an after-stay rope. The sails were linen, and each vessel had a single squaring sail that was furled by lowering the yardarm.

By 2500 B.C., Egyptians were building boats of wood. The method for building these boats was obviously inspired by the reed boat: Planks were sewn, edge to edge, without keel or ribs.

Metallurgy Technology

Sometime shortly before 3000 B.C., Egyptian metallurgists made a discovery that radically changed the metal product industry. They found that by mixing a small quantity of tin ore with copper ore during smelting, a harder and more useful metal, bronze, was obtained. Bronze was not only harder than copper; it was also more easily worked.

The first examples of bronze were found in the tombs of the early Sumerian kings. A study of the metal objects from the royal Sumarian tombs showed that the smiths were highly skilled. Complex objects were often cast in molds made of two to four pieces. It is very clear from the chemical composition of some weapons and ornaments that the smiths experimented widely with various alloys. They learned to join one piece of metal to another with an alloy of a different composition, a process commonly known as soldering (Hodges, 1970).

African-American Contributions to Technology

The aforementioned section illustrated *African* influences on technological development, but there were also many *African-American* contributions of record. As mentioned earlier, the technological creativity of Black slaves was repressed by the need to fulfill the economic needs of their respective plantation owners. Even then, slaves were never given proper recognition for many innovations because the patent belonged to the slave owner. Many innovations were developed to make slaves' lives and tasks a little easier. Post-slavery African-American technological innovations still did not enjoy proper recognition.

James Michael Brodie (1993) did an excellent job of compiling descriptions of the lives and ideas of African-American innovators. Because of his thorough research, we can synthesize and generate a matrix identifying and describing African-Americans' major contributions to engineering and engineering technology as they apply to technology education. The following matrix describes African-Americans and some of their contributions to technology based on the content areas that are identifiable to technology educators.

Retrospect and Prospect

African-Americans have had a profound effect on the development of technology and the technology education profession. At the beginning of this chapter, we briefly outlined how the enslavement and repression of Black African people by early colonists in the United States contributed to (and limited) the development of industrial and technological skills by Black people. Some technological developments occurred in early Africa, and knowledge of these technologies migrated from Africa to Europe (and eventually America) (Hilliard, 1990). The symbiotic relationship that Black people have with technology pre-dates their arrival to the United States as indentured servants.

Contributions of African-Americans to Technology Education

Table 1. African-Americans' Innovations and Technologies

Name	Content Area	Contribution	Year	Description
Benjamin Banneker	Surveying and Mapping	Design for the District of Columbia	1790s	First African to receive a presidential appointment
Henry Boyd	Construction	Boyd bedstead (bed frame)	1800s	Constructed the frame in such a way that its wooden bed rails could be screwed into both the headboard and the footboard, creating a strong structure
Thomas L. Jennings	Manufacturing	Dry cleaning process	1821	First African-American to receive a patent March 3, 1821
Norbert Rillieux	Bio-related	Sugar-refining process	1843	Developed a step-by-step process to turn heated evaporated sugar into crystallized granules
Lewis Howard Latimer	Electronics	Development of the electric light bulb	1848-1928	Only Black member of Thomas Edison's research team, developed and patented the process for manufacturing carbon filaments
Elijah McCoy	Manufacturing/ Transportation	Lubricator cup	1870s-1920s	Patented the cup in July 1872, to disperse small amounts of oil dripped onto moving engine parts, used with steam engines. By 1920, applied the technology to vehicles using air brakes
Granville T. Woods	Electricity/ Electronics	Telephone transmitter	1884	Improved clarity and sound of long distance signal
Jan Matzeliger	Manufacturing	Machine for attaching soles	1886	Created the only machine that could automatically attach soles to shoes. Required less than a minute to complete a single shoe

Table 1. (cont.)

Name	Content Area	Contribution	Year	Description
A. B. Blackburn	Transportation	Railway signal	1888	Developed a signal to alert automobiles, trucks buses, etc., of approaching trains
P. B. Downing	Transportation	Electric switch for railways	1890	Developed a switch to enable the automatic switching of train tracks as opposed to manual switching
Andrew Jackson Beard	Transportation	Rotary engine	1892	Developed a more efficient rotary engine for automobiles and trucks
Joseph Lee	Bio-related Technology/Manufacturing	Bread-crumbling machine/Bread-making machine	1895	Reduced bread to crumbs by tearing and grinding, a more sanitary way to knead dough
Andrew Jackson Beard	Transportation	Automatic railroad coupler (Jenny Coupler)	1897	Allowed trains to be joined without human assistance
George Washington Carver	Bio-related	Products derived from sweet potato, peanuts, pecans, and clay	1909	Developed more than 300 products derived from peanuts, 100 from sweet potatoes, about 75 from pecans, and many more from Georgia clay
Garrett Augustus Morgan	Transportation	Automatic traffic signal	1923	Helped revolutionize traffic control and led to the later development of the overhead signal
Charles Richard Drew	Bio-related Technology	Blood bank	1935	Devised a way to separate plasma from whole blood, making the blood easier to store and paving the way for the creation of the first blood banks

Following slavery, African-American people continued their involvement in the development and utilization of technology and technological systems. Even though many developments improved life for *all* Americans, they went virtually unrecognized in American history.

African-Americans and Technology Education: A Prospectus

Technology education educators face an important challenge—to infuse technology education into all educational programs for *all* students. Issues at all levels of the educational system, from elementary schools through the universities, reflect an unstated belief that perhaps technology education is *not* appropriate for all students. Although technology education programs are flourishing in suburban school districts in America, there is little evidence that the same is true in the largest urban school districts. Why, for example, are technology education programs in these districts not represented in program and teacher-of-the-year recognitions at state and national levels? Why is there also a dearth of African-American teachers in these programs?

Although segregation was outlawed in the United States, Anderson (1982) stated that African-American and White systems of industrial education continued to operate well into the 1960s. Historically Black Colleges and Universities (HBCUs) offered opportunities for African-American students to earn degrees in technology education when they could not attend "non-Black" institutions (Johnson, 1996). Still, African-Americans are not well represented in technology education programs (Hall, 1973; Lewis, 1991; Wolfson, 1986).

Doctoral programs in technology education have produced few minority graduates, and most historically non-Black institutions have no Black faculty. A few institutions have a small number of minority faculty (women or people of color). Some minority graduates choose to move beyond a faculty position into administration within the university or move out of the educational arena entirely. Recruitment and retention into teacher preparation programs as well as graduate programs is essential to the diversification of the profession.

The technology education profession has begun to show some signs of progress. For example, the International Technology Education Association presents the Rutherford E. Lockette Humanitarian Award annually. This award honors Rutherford E. Lockette, an African-American teacher educator in the 1960s and 1970s. In the mid-1990s, the membership of the International Technology Education Association elected its first African-American president, John Monroe, a teacher from Virginia.

The International Technology Education Association, the National Science Foundation (NSF), and the National Aeronautics and Space Administration (NASA) recently commissioned the *Technology for All Americans Project* (International Technology Education Association, 1996). By the project's end, we hope that this effort will truly, as its title states, provide leadership for the study of technology for *all* Americans.

REFERENCES

Anderson, J. D. (1982). The historical development of black vocational education. In H. Kantor & D. Tyack (Eds.), *Work, youth and schooling* (pp. 180-222). Stanford, CA: Stanford University Press.

Anderson, J. D. (1988). *The education of Blacks in the south.* Chapel Hill: University of North Carolina Press.

Barlow, M. L. (1967). *History of industrial education in the United States.* Peoria, IL: Chas A. Bennett.

Bennett, C. A. (1926). *History of manual and industrial education up to 1870.* Peoria, IL: Author.

Black history at an early age: A gift of heritage African kings and queens (Vol. 6) (1991). Chicago: An Empak "Black History" Publication Series.

Brodie, J. M. (1993). *Created equal: The lives and ideas of black American innovators.* New York: Morrow.

Brooks, A. S., Helgren, D. M., Cramer, J. S., Franklin, A., Hornyak, W., Keating, J. M., Klein, R. G., Rink, W. J., Schwarcz, H., Leith Smith, J. N., Stewart, K., Todd, N. E., Verniers, J., & Yellen, J. E. (1995). *Dating and context of three middle stone age sites with bone points in the Upper Semliki Valley, Zaire,* 268(5210), 548.

DuBois, W. E. B. (1932). *Education and work.* Journal of Negro Education, 1(1), 60-74.

Hall, C. W. (1973). *Black vocational technical and industrial arts education development and history.* Chicago: American Technical Society.

Hayden, R. C. (1972). *Eight Black American inventors.* Reading, MA: Addison-Wesley.

Hilliard, A. (1990). *Teacher education and the African American student.* Unpublished paper, The Ohio State University.

Hodges, H. (1970). *Technology in the ancient world.* London, WI: Penguin.

International Technology Education Association. (1996). *Technology for all Americans.* Reston, VA: Author.

Johnson, K. (1996). Some thoughts on African Americans' struggle to participate in technology education. *Journal of Technology Studies*, XXII(1), 49-54.

Lewis, T. (1991). Main currents in teacher education: Imperatives for technology teacher education. *Journal of Industrial Teacher Education*, 1(29), 25-45.

Lux, D. G. (1981, April). Reality, Aristotle, and the teaching of learning about and learning how-to. *School Shop*, 40(8), 24-25.

Mikami, T. (1972). *The art of Japanese ceramics*. New York: Weatherhill/Heibonsha.

Moody, F. (1980, January). The history of Blacks in vocational education. *Vocational Education Journal*, 30-34.

Paulsen, G. (Speaker) (1993). *Nightjohn* (Cassette Recording). Prince Frederick, MD: Recorded Books.

Spivey, D. (1978). *Schooling for the new slavery. Black industrial education, 1868-1915*. Westport, CT: Greenwood.

Wolfson, R. A. (1986). *Job satisfaction of industrial arts/technology teacher education faculty in the United States*. Unpublished doctoral dissertation, The Ohio State University

Yellen, J. E., Brooks, A. S., Cornelissen, E., Mehlman, M. J., & Stewart, K. (1995). *A middle stone age worked bone industry from Katanda, Upper Semliki Valley, Zaire*, 268(5210), 553.

Section II: UNDERREPRESENTED GROUPS AS TECHNOLOGY STUDENTS AND EDUCATORS

Women as Technology Educators

Colleen E. Hill
California State University, Long Beach

AN OCCUPATION THAT HAS NO BASIS IN SEX-DETERMINED GIFTS CAN NOW RECRUIT ITS RANKS FROM TWICE AS MANY POTENTIAL ARTISTS.
(Margaret Mead, 1935)

Technological literacy, as a national imperative, is critical for all students. *Technology For All Americans* (Satchwell & Dugger, 1995) is the latest national effort to create structure for the field. One would assume that *All Americans* means that the technology education profession is inclusive in terms of gender, race, ethnicity, and ability. This means that attention needs to be paid to making the field responsive to the needs of all students. However, there are very few females in technology education, and a lack of diversity in any arena in the 1990s is problematic (Bame & Dugger, 1990; Erekson & Trautman, 1995a,b; Heidari, 1994; Liedtke, 1986; Markert, 1981, 1996b; Miller, 1993; Rider, 1991; Silverman & Pritchard, 1993; Trautman, Hayden, & Smink, 1995; Zuga, 1992, 1996a). If technology for all Americans is to become reality, more women must be encouraged to enter and be supported within technology education.

Welty (1996) stated that "(t)he absence of women among the ranks of technology educators reinforces the subliminal misconception that the study of technology education is a male endeavor" (p. 2). Women in technology education also may be somewhat skeptical of the claim that research on women's issues is being done in the male-dominated field. Thus, it is time that we all join forces to make the profession a better place to study and work. As Heilman and Goodman (1996, p. 254) noted, "(b)oth females and males need to appreciate issues of gender as one of many human social problems and participate in its resolutions."

Women's struggle for legitimacy needs to be addressed by both genders, but not all individuals recognize that it truly is a struggle.

The struggle includes the attempt for females to achieve success; gender discrimination and stereotyping; family/societal pressures and lack of encouragement/support; recruitment, retention, and advancement of women in the field; and, finally, the attempt to move technology education from a field that is relatively homogeneous to a diverse and dynamic field that includes all and truly is "technology for all Americans."

ACHIEVING SUCCESS

To achieve success in any endeavor, one must have qualities that enable him or her to "go for it." Fortunately, those qualities are not related to gender, culture, race, or physical trait. Markert (1996a) wrote that "(r)egardless of the discipline, successful people get ahead because of their intellect, energy, enthusiasm, determination, character, communication skills, charisma, and a modicum of luck! None of these traits is a gender-linked phenomenon" (p. 6). Other authors have come to similar conclusions when examining success. "Women who persevere . . . often have similar qualities, including a sense of confidence, a high self-esteem, a strong locus of control, a firm desire to succeed, a refusal to be limited by their gender, and a full capability in the fields they pursue" (Brusic, 1996, p. 6). Moss and Jensrud (1995) found all leaders are perceived to have the same leadership attributes; leaders are seen as "energetic with stamina . . . visionary . . . confident, accepting of self, willing to accept responsibility, persistent, enthusiastic, optimistic . . . Men and women who are recognized as leaders by their colleagues are perceived to have the same set of leader attributes" (p. 12).

In a survey of successful men and women leaders in her corporation, Angelo (1996) asked what contributes to women's success. Answers to the question included native intelligence, political and communication skills, self confidence/assertiveness, genuine interest in business issues, superior active listening skills, the ability to perceive interrelationships/interdependencies among issues, and sensitivity to emotional/intuitive factors in decision making. Of all these traits for achieving success, none was gender specific.

Nevertheless, barriers affect the success of females in traditionally male-dominated fields. These barriers need to be identified so that they can be circumnavigated at least and torn down at best. What are the barriers that preclude women from gaining access to technology education and achieving success? A review of the literature in technology education and other fields revealed a plethora of issues that illustrate these barriers.

GENDER DISCRIMINATION AND STEREOTYPING

Daniels' (1996) statement that "negative and often inaccurate stereotypes associated with technology may discourage women from seriously considering careers in engineering" (p. 4) could be read as "careers in technology education" as well. A female technology teacher educator told this story of being interviewed for a teaching position. After she accepted the job, several of her future male colleagues apologized for voting against her being hired. Although they felt that she was the best candidate, they did not want to be coerced into hiring a woman just to comply with pressure to diversify their all-male faculty. Gender obviously came into play here, although this woman got the appointment. Gender discrimination remains a very real concern to women seeking entrance into the ranks of technology teacher educators.

What stereotypes do women face? Belenky, Clinchy, Goldberger, and Krule (1986) explained: "It is likely that the commonly accepted stereotype of women's thinking as emotional, intuitive and personalized has contributed to the devaluation of women's minds and contributions, particularly in Western technologically oriented cultures, which value rationalism and objectivity" (p. 6). Gilligan (1982) elaborated that "these differences arise in a social context where factors of social status and power combine with reproductive biology to shape the experience of males and females and the relations between the sexes" (p. 2). Another writer concurred that "(g)ender is not just about difference but about power . . . technical competence is central to the ideal of masculinity and its absence is a key feature of stereotypical femininity" (Wajcman, 1991, p. 159).

Technology education, like the profession, has members in all levels who, consciously or unconsciously, exhibit stereotypical behaviors. These behaviors must be examined and dislodged. Grossman and Grossman (1994) stated that "if you wish to eliminate your students' stereotypical behavior, first examine your own behavior for possible gender bias and correct any that you may discover" (p. 167). The elimination of behaviors, including language, that contribute to discriminatory practice aids in the movement toward a diverse culture. Liedtke (1996) explained that "(l)anguage systems and metaphors used in an organization also indicate the shared values. How people refer to each other in professional settings and the language used can create the feeling of an open group or closed society" (p. 3).

It is difficult for women to "be productive in environments that lower their self-esteem, isolate them, and minimize their contributions" (Brusic, 1996, p. 4). Unfortunately, women's contributions in technology often have been ignored or lost (Wajcman, 1991; Zuga, 1996a,b), adding to the notion that women are devalued or not valued at all in the discipline. Liedtke (1995) explained that "people usually accept (or gravitate toward) organizational cultures that are like themselves and where they can feel comfortable and contribute" (p. 11).

Although technology education professionals cannot eliminate all commonly held stereotypes concerning women, they can work toward creating a gender-neutral environment within the discipline. To do that, support networks that include men and women need to be established, and an aggressive plan for increasing diversity within the field needs to be formulated and implemented.

SUPPORT/ENCOURAGEMENT AND FAMILY/OUTSIDE PRESSURES

Because of stereotyping, women too often are subjected to pressures of family responsibilities, pressures as a result of the belief structure of society, and lack of support and/or encouragement in choosing a non-traditional occupation or career. A woman who teaches high school technology expressed concern that if she pursues her doctorate she will no longer be attractive to men and that in doing so she will be giving up the chance for marriage and family. Although this is probably an unfounded fear, it is still very real to her. Unfortunately, earning a degree or working in higher education too often is still seen as a detriment to a woman's future role as wife and mother. This attitude or belief leads one to assume that gender and achievement conflict. Some studies have indicated that "depression, infertility and reduced marriage prospects of working women" (Heilman & Goodman, 1996, p. 250) perpetuate the myth that women's careers are not as important as men's.

Most people have a relatively clear idea of what type of work individuals, usually according to their gender, ethnicity, or ability, can and should be doing. Traditionally, women have been relegated to lower status, lower paying positions. Many people believe that because females biologically give birth, sociologically they are responsible for child care and family maintenance. Hartman (1987) spoke of "the uneven responsibility and rewards of the two sexes in family life" (p. 110). She also suggested that "the sexual division of labor by gender makes men primarily responsible for wage labor and women primarily responsible for household production" (p. 114). Tavris (1993) called it the "'second shift': housework, child care, family obligations" (p. 296). Most women are responsible for unpaid domestic work and child care (Foster, 1996), and this caring, nurturance, and emotional support often create an extreme burden that competes with the burden of career. Hence, the supermom issue and the juggling of work, family, and self creates a barrier to career and personal success.

Angelo (1996) surveyed men and women leaders in her corporation concerning women's contributions in technology professions. She found that men thought that reasons women "derail" from a career path included "conflicts between the demand of family and career, frustration with having to defer to superiors who don't take them seriously, (and a) lack of (an on-the-job) support system" (p. 7). When Angelo asked women about the unique challenges that they faced, their responses included "women (have) to be constantly proving themselves to gain respect . . . Overcoming stereotypes, lack of role models and managing family and career were also frequently cited" (p. 5).

Fortunately, through laws and institutional or corporate policies, outright discrimination is on the wane overall. However, sometimes there is a covert lack of support that is hard to fight. It is the subtle sabotage that happens through withheld enthusiasm and often keeps women from making connections and having professional opportunities to advance. These types of support are important to success in the workplace. Women in non-traditional fields often feel that they are outsiders and choose to leave the profession rather than fight the status quo, which can cause perceived feelings of inadequacy and failure.

Brusic (1996) stated that "(i)nsufficient support networks exist for women to help them overcome the obstacles they face in their technological endeavors and too often they give up rather than fight a seemingly exhaustive battle" (p. 5). More optimistically, she wrote that "perhaps these hurdles will eventually disappear as men and women learn to respect their differences and empower each other to succeed" (p. 6). A concerted effort at all levels needs to be made so that men and women learn to respect their differences and support each other's efforts in the technology education profession. Starkweather (1996) acknowledged the following:

> Engaging women in the conduct and decision making of technology teaching needs to be supported and encouraged until this issue is no longer a concern. When this goal is achieved, technology teaching will be enhanced, and the technology profession, as a whole, will benefit greatly, because it will finally be utilizing the full range of human resources in our society. (p. 1)

Liedtke (1996) cautioned us to look deeper and explained that without thorough examination of all facts, a common reaction to high turnover among women is to improve the benefits package, because it is believed that women leave the workplace to start families. She explained that the truth is that women leave more often because they do not relate to the organization and see limited opportunities for satisfaction in their chosen work. Brown and Gilligan (1992) stated that in working with girls and women, professionals need to be "particularly attentive to their struggles for relationships that are authentic or resonant, that is, relationships in which they can freely express themselves or speak their feelings and thoughts and be heard" (p. 29). To not do so will perpetuate the lack of encouragement and support within the field, making it much more difficult to attract women to technology education.

RECRUITMENT, RETENTION, AND ADVANCEMENT

Women have always been seriously underrepresented in all levels of technology education. This issue parallels a growing concern within the profession of a severe and growing shortage of new technology teachers to fill positions in middle and high school programs and a coinciding shortage of technology teacher education programs and faculty (Lewis, 1991; Miller, 1991; Moss, 1989; Oaks & Loepp, 1989; Scarborough, 1990; Scarcella, 1997; Starkweather, 1996; Volk, 1993; Welty, 1996; Wicklein, 1993; Wright, 1991).

Unsupportive administrators outside the technology profession also hold back female technology education teachers or technology teacher educators. For example, administrators may determine what is "appropriate" for women to teach. One middle school technology education teacher, who holds certification in several subjects, was advised by school administrators that she should go back to teaching English. She was told that English is a more necessary part of the curriculum and more adequately fits with what women are supposed to teach. This woman is still teaching technology education, but subtle changes in the teaching assignments have forced her into at least partial compliance with this practice. She feels she has no options or choice in the matter.

We should not lose good people from our field, especially if they can help increase diversity in the profession. Welty explained that recruitment, retention, and advancement of women in "technology education can help break the cycle of events that are discouraging girls and young women from entertaining and pursuing non-traditional careers" (1996, p. 12). Instead of hindering women who pursue a career in our profession, we must begin to encourage and support them.

Salinger (1994) challenged the profession with this question: Are we recruiting for technology education or for technology educators? Tracey (1996) indicated that the primary incentive to increase the number of female faculty in higher education is not just to comply with Affirmative Action requirements but to add diversity and increase the number of role models and mentors to continue the trend toward diversity.

Another question that parallels Salinger's challenge is this: Are we trying to recruit women into the field because this population has the potential to increase significantly the numbers of students in our classes and we need jobs as technology educators or because technological literacy is not (or at least should not be) a gender-linked phenomenon? The need to address recruitment of females into technology education, as illustrated in the International Technology Education Association's 1993-1995 strategic planning goals, is to enhance participation of minorities and women in technology (Householder, 1993; Starkweather, 1995). We need to ensure that our recruitment motives are pure and not self-serving.

One method of recruiting, retaining, and promoting women in technology education is to establish role models and mentors (Daniels, 1996; Householder, 1996; Liedtke, 1986; Starkweather, 1996; Tracey, 1996; Welty, 1996). Young-Hawkins (1996) explained that "mentoring has become widely recognized as an important source of development for individuals . . . (and that) mentors are helpful for white males, but essential for women and for ethnic minorities" (p. 2). Welty (1996) concurred: "Without female role models, both within public education as well as from the private sector, girls and young women will be reluctant to consider the study of technology and the pursuit of non-traditional occupations among the options" (p. 2).

Daniels (1996) encouraged the use of role models but cautioned that "unless role models are particularly adept at talking to children or teenagers about their career, they may discourage and confuse the audience with their technical vocabulary or overwhelming accomplishments" (p. 4). Welty, however, also warned that "some experts recommend avoiding placing too much emphasis on the 'heroines of technology'" (1996, p. 3). Householder (1996) further commented that

> The study of high achievers may not be the most fertile way to develop a strategy for attracting more women to technology and science. From what we know about the appropriateness of the mentor-protégé relationship, it may be that moderately successful individuals communicate a more realistic expectation for attainment. (p. 3)

Liedtke (1986) suggested that a developmental group for potential mentors should be instituted, perhaps through professional associations. This group could encourage and teach mentoring strategies and help make significant connections. Starkweather (1996) concurred: "We must not be satisfied with an informal happenstance mentoring system but must offer a full program of support that is known, utilized, and is self perpetuating" (p. 7).

Summer camps have been used as recruitment tools to attract female high school students into technical or technology fields (Husher, 1993; Volk & Holsey, 1997). Daniels (1996) described one successful recruiting strategy: A hands-on lab for first-year women students at Purdue University encouraged them to consider engineering. According to Daniels, "the experience developed feelings of familiarity and self-confidence in the students" (p. 5).

Liedtke (1996) cautioned that recruitment will only be beneficial if retention and advancement issues are addressed. Householder (1996) suggested that networking is an important part of retention and advancement. As Young-Hawkins (1996, p. 4) stated, "(c)learly, there is a need to develop and implement a plan of action that supports and engages women in the conduct of the profession, increasing access to leadership opportunities, administrative positions, and improving the climate for women in technology education at all levels." She further wrote that "(t)he ultimate goal . . . is to set in motion an ongoing process for support systems and mentors in the future to enable women to remain in the pipeline" (Young-Hawkins, 1996, p. 5). Israel (1996) advocated that women "be provided the opportunity to assume successful leadership roles and become involved in the major decisions in professional associations and in their chosen profession" (p. 3).

Becoming a Mosaic

In the early 1980s, a wonderful social studies program was taught in junior high schools in Calgary, Alberta. It was about society's mosaic and how, although each tile is different, uniqueness must be celebrated as it adds to the strength of the portrait. Today we need to rethink the structure of the technology education profession to create a mosaic that celebrates diversity.

Zuga (1996b) stated that "there are differences in the way in which women experience and, therefore, think about the world" (p. 10). She elaborated that "(i)t appears as though the problem is much deeper than socializing girls and women into the existing hierarchy. Perhaps, we need to rethink the hierarchy and technology education content and practice" (p. 11).

The need to celebrate diversity also exists in other related fields. Alper (1993, p. 411) noted that "it's hard to escape the conclusion that the culture of science itself may also have something to do with women's lack of interest in pursuing a scientific career." We must move beyond trying to equip women to survive in traditionally male-centered fields such as technology education and instead equip technology education to be responsive to the needs of men and women in the profession.

Teaching women to succeed in the masculine environment of technology education is not the same as developing a non-sexist or even anti-sexist environment. What, then, constitutes an environment free of discrimination? A lesson might be learned from the special education field, where there has been pressure to shift paradigms to redefine the meaning of disability in the social/cultural context of the 1990s. A comparison of the predominant two paradigms can be used to redefine the meaning of diversity in technology education.

The individual defect paradigm is focused on the individual with a disability. The problem is the individual's failure to perform major life activities, and the solutions sought are to restore an individual's function and help him or her adapt to the defects. Strategies used to support the individual include medical evaluation and treatment, rehabilitation, and services designed for and used exclusively by people with disabilities. The consequences are that the individual gains in functional abilities, acquired trade skills, and acceptance of and reliance on service systems. The desired outcomes are improved functional capacity, improved personal adjustment, and less use of support services.

The second paradigm is concerned with the individual rights of the person with a disability. The focus is on systems, laws, and relationships; the nature of the problem is persistent and pervasive, but often unconscious, discrimination; and the solutions sought are access to society's economic, social, educational, and environmental tools. The strategies here are nonviolent confrontation, civil rights legislation, policy reform, consciousness raising, and political action. The consequences of these strategies are the removal of systemic and structural barriers; regarding oneself and being regarded as an equal citizen; and greater reliance on the economic, environmental, legal, and social tools used and valued by mainstream America. The desired outcomes of this paradigm are equal opportunity, freedom of choice, and the rights and responsibilities of full citizenship.

The difference between the two paradigms is that the first is focused on fixing the problems of the individual and the second is focused on fixing the problems of the society within which the individual has to function. These foci can be related to the process of the assimilation of diverse newcomers, who are expected to do all the changing and adapting to the organization. However, the organization needs to change and adapt to accommodate diverse individuals if it truly wants to include them. Liedtke (1996) wrote that it doesn't work to recruit women and then fix their behavior. Fixing the behavior of those who are different to make them more like the established norm results in a lack of diversity and a flood of underrepresented groups leaving the organization.

The experiences of women must be revalued; however, care must be taken not to condone characteristics that have been developed in response to male domination or to denigrate male characteristics en route. A new synthesis needs to be created for everyone, allowing both male and female experiences to be seen, valued, and rethought (Gaskell & McLaren, 1991). Liedtke (1995) advised the following: "To increase the participation of women in technology education as a profession, there must be an organizational culture which is attractive to these individuals and is consistent with the factors (values and norms) which these individuals can best identify" (p. 9). To do this, technology organizations such as educational institutions, professional associations, and business/industry must develop and reinforce organizational cultures that ensure diversity and value women.

According to Gaskell and McLaren (1991), gender is fundamental to the ways we interact with each other and the ways our public and private lives are organized. They also explained that

> The feminist critique has no single voice. Some feminists want to add women's concerns to an existing curriculum; others want to reshape the entire curriculum. Some believe there is a distinctive women's way of discovering and knowing the world; others want to focus on the differences among women. Some focus on the content of what is taught; others focus on how it is taught, arguing that the medium is the message. Some women are frustrated and angry at the recalcitrance of those who refuse to respond to women's concerns about curriculum; others are excited and optimistic about the potential for reviewing and revitalizing education which the discovery of women's choices brings. In all cases, feminists critically examine what counts as education and want it changed. (p. 225)

According to many women in male-dominated fields, to be more attractive to females and to echo their voices, the curriculum needs to be more humanistic and environmental (O'Riley, 1996; Zuga, 1992, 1996b). Krockover and Shepardson (1995, p. 223) argued that "(p)rospective teachers must be able to implement gender equitable curricula, instructional materials, and assessment strategies. Can we afford anything less for the greater than 50 percent of our global population that has been educationally disenfranchised for centuries?" It is not enough to bandage the problem by placing girls in technology education classrooms, making sure that the graphics in texts include females, or using gender-free language. Although these bandages help, they mask the deeper problem of the structure and stricture of power (Stone, 1996). We must make a compelling argument for a gender-friendly technology education curriculum and classroom. As Zuga noted, "(i)mplementing a social reconstruction curriculum design in technology education encourages thoughtful critique of the status quo and existing practice with respect to technology" (1996b, p. 13).

FINDING EQUITABLE SOLUTIONS

Like design problems that are addressed through the technological problem-solving method, multiple solutions must be used to address the variety of gender-inequitable situations. There is no shortage of roadmaps on what needs to be done. The avenues are there; the people who are interested in advancing solutions need to help others willing to travel the road. Members of our profession need to take the lead in formulating a plan of action, ensuring diversity of membership, and integrating the profession from the onset. Starkweather (1996) stated that

> New initiatives are imperative to advance women in technology teaching positions, whether elementary, secondary, college/university or at the corporate level. Action is needed at all levels to advance more women into leadership roles. Strategic plans to implement action for the advancement of women in technology teaching are long overdue. (p. 3)

We must heed the actions of other groups in which similar gender-based underrepresentation exists. Martin (1996) stated that "(t)here is a growing list of resources available to both genders on ways to increase and capitalize on access, but little has been published in the field of technology" (p. 1). We can adapt some lessons learned in science and engineering, for example, as strategies for combating gender discrimination and stereotyping. Daniels (1996) encouraged us to include parents in our education plan to dispel stereotypes: "Parents of young women thinking about engineering as a career choice are frequently concerned that their daughter may be choosing a lifestyle which is incompatible with having a successful marriage or family—one in which they must behave in masculine ways" (p. 5). By including parents in our mission to diversify the profession, we can help allay their fears.

We need support within the ranks of the profession and from outside it to recognize technology education as a school subject that is appropriate (and necessary) for all students. The profession needs to create and actively implement a public relations and recruitment plan. Isbell and Lovedahl (1989) indicated that "recruiting can be successful if faculty use a well-planned, professional approach, involving a team effort among teacher educators, supervisors, counselors, and classroom teachers" (p. 41). However, the "add women and stir" agenda will not make the profession more gender friendly. We will know we have been successful when, as Murphy (1996) stated, "our daughters are inspired to follow us into technical careers because our jobs appear to be so interesting and personally rewarding" (p. 7). It will be at that point that, we hope, we will have gotten past the frustrations of being "women in" technology education. Nonetheless, until the whole profession is re-examined, re-valued, re-visioned, and the interests of All are attended to, it will remain business as usual, and Technology For All Americans will be a misnomer for the standards of our profession.

REFERENCES

Alper, J. (1993, April 16). *The pipeline is leaking women all the way along.* Science, 260, 409-411.

Angelo, S. (1996, October). *The contributions of women in technology professions: A business and industry perspective.* Paper presented at the Women's Leadership Symposium, Chicago.

Bame, E. A., & Dugger, W. E., Jr. (1990). Pupil's attitudes and concepts of technology. *The Technology Teacher,* 49(8), 10-11.

Belenky, M. F., Clinchy, B. M., Goldberger, N. R., & Tarule, J. M. (1986). *Women's ways of knowing: The development of self, voice, and mind.* New York: Basic Books.

Brown, L. M., & Gilligan, C. (1992). *Meeting at the crossroads.* Cambridge, MA: Harvard University Press.

Brusic, S. A. (1996, October). *The path of least resistance: The retention of women in technological endeavors.* Paper presented at the Women's Leadership Symposium, Chicago.

Daniels, J. Z. (1996, October). *Purdue's women in engineering program: Recruitment and retention strategies.* Paper presented at the Women's Leadership Symposium, Chicago.

Erekson, T. L., & Trautman, D. K. (1995a). Diversity or conformity? *Journal of Industrial Teacher Education,* 32(4), 32-42.

Erekson, T. L., & Trautman, D. K. (1995b). Cultural diversity and the professions in technology. *The Journal of Technology Studies,* XXI(2), 36-42.

Foster, V. (1996). Space invaders: Desire and threat in the schooling of girls. *The High School Journal,* 79(3), 191-201.

Gaskell, J. S., & McLaren, A. T. (Eds.). (1991). *Women and education* (2nd ed.). Calgary, AB: Detselig Enterprises.

Gilligan, C. (1982). *In a different voice: Psychological theory and women's development.* Cambridge, MA: Harvard University Press.

Grossman, H., & Grossman, S. H. (1994). *Gender issues in education.* Boston: Allyn and Bacon.

Hartman, H. I. (1987). The family as the locus of gender, class, and political struggle: The example of housework. In S. Harding (Ed.), *Feminism and methodology* (pp. 109-134). Indianapolis: Indiana University Press.

Heidari, F. (1994). *Demographic survey of female faculty in technology education programs* (Report No. CE-067-419). Eastern New Mexico University. Paper submitted to Educational Resources Information Center. (Eric Document Reproduction Service No. ED 375 275)

Heilman, E., & Goodman, J. (1996). Teaching gender identity in high school. *The High School Journal, 79*(3), 249-261.

Householder, D. L. (1993). Creating the future: Strategic planning and organizational change and ITEA strategic plan. *The Technology Teacher, 52*(7), 3-8.

Householder, D. L. (1996, October). *Providing role models: The importance of mentors for women in technology.* Paper presented at the Women's Leadership Symposium, Chicago.

Husher, H. (1993, February). Closing the gap—Women in technology. *Tech Directions,* 15-17.

Isbell, C. H., & Lovedahl, G. G. (1989). A survey of recruitment techniques used in industrial arts/technology education programs. *The Journal of Epsilon Pi Tau, XV*(1), 37-41.

Israel, E. N. (1996, October). *Engaging women in the conduct and decision making on the Council on Technology Teacher Education (CTTE) and the National Association of Industrial Technology (NAIT).* Paper presented at the Women's Leadership Symposium, Chicago.

Krockover, G. H., & Shepardson, D. P. (1995). Editorial: The missing links in gender equity research. *Journal of Research in Science Teaching, 32*(3), 223-224.

Lewis, T. (1991). Main currents in teacher education: Imperatives for technology teacher education. *Journal of Industrial Teacher Education, 29*(1), 25-52.

Liedtke, J. A. (1986). Mentors and role models: Influences on the professional career. *The Journal of Epsilon Pi Tau, XII*(1), 41-44.

Liedtke, J. (1995). Changing the organizational culture of technology education to attract minorities and women. *The Technology Teacher, 51*(6), 9-14.

Liedtke, J. A. (1996, October). *How professional cultures impact women.* Paper presented at the Women's Leadership Symposium, Chicago.

Markert, L. R. (1981, October). Women researchers in science and technology: Why so few? *Man, Society, Technology, 41*(1), 12-14.

Markert, L. R. (1996a). Gender related to success in science and technology. *Journal of Technology Studies,* XXII(2), 21-29.

Markert, L. R. (1996b, October). *Increasing access to leadership in technology: Developing a national agenda and support system.* Paper presented at the Women's Leadership Symposium, Chicago.

Martin, G.E. (1996, October). *Increasing access to leadership opportunities and administrative positions in technology.* Paper presented at the Women's Leadership Symposium, Chicago.

Miller, J. A. (1991). Recruitment and support for women students in technology teacher education. *The Journal of Epsilon Pi Tau,* XVII(2), 27-30.

Miller, J. L. (1993). Constructions of curriculum and gender. In S. K. Biklen & D. Pollard (Eds.), *Gender and education* (pp. 43-63). Chicago: University of Chicago.

Moss, J., Jr. (1989). Contemporary challenges for industrial teacher education. *Journal of Industrial Teacher Education, 26*(2), 23-28.

Moss, J., Jr., & Jensrud, Q. (1995). Gender, leadership, and vocational education. *Journal of Industrial Teacher Education, 33*(1), 6-23.

Murphy, M. A. (1996, October). *The importance of women's leadership in technology professions.* Paper presented at the Women's Leadership Symposium, Chicago.

Oaks, M., & Loepp, F. (1989). The future of technology teacher education. *Journal of Industrial Teacher Education, 26*(4), 67-70.

O'Riley, P. (1996). A different storytelling of technology education curriculum re-visions: A storytelling of difference. *Journal of Technology Education, 7*(2), 28-40.

Rider, B. L. (1991). *Problems and issues facing women in technology education*. Paper presented at the Mississippi Valley Industrial Teacher Education, Nashville, TN.

Salinger, G. L. (1994, November). *Research and marketing issues for technology education*. Paper presented at the Mississippi Valley Industrial Teacher Education Conference, Chicago.

Satchwell, R., & Dugger, W. E., Jr. (1995). Our challenge: Technology for all Americans. *Journal of Industrial Teacher Education, 32*(2), 93-94.

Scarborough, J. D. (1990). Personal and professional needs of technology teachers. *Journal of Teacher Education, 1*(2), 42-51.

Scarcella, J. A. (1997). Technology education recruitment: Challenges for the 21st century. *The Technology Teacher, 56*(4), 2-3.

Silverman, S., & Pritchard, A. M. (1993). *Guidance, gender equity, and technology education*. Hartford, CT: Vocational Equity Research, Training and Evaluation Center.

Starkweather, K. N. (1995). The International Technology Education Association. In G. E. Martin (Ed.), *Foundations of technology education* (pp. 543-566). New York: Glencoe.

Starkweather, K. N. (1996, October). *Engaging women in the conduct & decision making of technology professions*. Paper presented at the Women's Leadership Symposium, Chicago.

Stone, L. (1996). Disruptive teaching: An introduction to the special issue. *The High School Journal, 79*(3), 167-175.

Tavris, C. (1993). *The mismeasure of woman*. New York: Simon & Schuster.

Tracey, K. C. (1996, October). *Advancing women's leadership in technology: Research in technology education*. Paper presented at the Women's Leadership Symposium, Chicago.

Trautman, D. K., Hayden, T. E., & Smink, J. M. (1995). Women surviving in technology education: What does it take? *The Technology Teacher, 54*(5), 39-42.

Volk, K. (1993). Enrollment trends in industrial arts/technology teacher education from 1970-1990. *Journal of Technology Education, 4*(2), 46-59.

Volk, K., & Holsey, L. (1997). TAP: A gender equity program in high technology. *The Technology Teacher, 56*(4), 10-13.

Wajcman, J. (1991). *Feminism confronts technology.* University Park: The Pennsylvania State University Press.

Welty, K. (1996, October). *Providing role models and mentors for women in technology.* Paper presented at the Women's Leadership Symposium, Chicago.

Wicklein, R. C. (1993). Identifying critical issues and problems in technology education using a modified-Delphi technique. *Journal of Technology Education, 5*(1), 54-71.

Wright, M. D. (1991). Retaining teachers in technology education: Probable causes, possible solution. *Journal of Technology Education, 3*(1), 55-69.

Young-Hawkins, L. (1996, October). *Mentoring women in technology: Action for advancing women's leadership.* Paper presented at the Women's Leadership Symposium, Chicago.

Zuga, K. F. (1992). Social reconstruction curriculum and technology education. *Journal of Technology Education, 3*(2), 53-63.

Zuga, K. F. (1996a). Reclaiming the voices of female and elementary school educators in technology education. *Journal of Industrial Teacher Education, 33*(3), 23-43.

Zuga, K. F. (1996b, October). *Women's ways of knowing and technology education.* Paper presented at the Women's Leadership Symposium, Chicago.

Chapter 5: Minority Students

Elazer J. Barnette
North Carolina A&T State University

THE ABSENCE OR LACK OF ROLE MODELS FOR MINORITY STUDENTS WOULD RESULT IN EDUCATIONAL DEFICITS FOR THE NATION'S YOUTH. SUCH A DEFICIT WOULD THREATEN AMERICA'S FUTURE PROSPERITY AND ABILITY TO COMPETE WHEN COMPARED TO OTHER INDUSTRIALIZED NATIONS OF THE WORLD.
(American Association of Colleges, 1988, p. 13)

While sitting at the 1988 annual Epsilon Pi Tau (EPT) breakfast in Norfolk, VA, I saw the leadership of the International Technology Education Association (ITEA) and realized that something was missing. That something was the lack of women and minorities in EPT and ITEA leadership roles. A major goal established at that breakfast was for the leadership and membership of ITEA to become more diversified through increased participation of women and minorities in the professional association.

However, eight years later, while sitting at the 1996 EPT breakfast in Phoenix, AZ, I noted that the face of the profession looks very much as it did in 1988. What happened? Are there no women or minorities in technology education who want to become involved in leadership roles? Are women and minority technology teachers so disillusioned with the lack of representation in the association that they choose not to participate and are not encouraging minority students to enter technology education? If we want the profession to change, minority students' participation in technology education and technology leadership activities must be encouraged. Those in leadership roles in ITEA and its affiliate councils should move forward to establish action plans to alleviate the situation. I offer suggestions to assist in establishing a pipeline for minority students who are equipped with the requisite leadership skills that can position them for leadership roles.

MINORITY TECHNOLOGY STUDENTS: WHO AND WHERE ARE THEY?

When one hears the words *minority students*, the immediate perception is of Black students or a Black group. From a narrow perspective, that perception has validity, but it only indicates one minority group. Because minority groups share many common concerns, people of color are frequently classified as a single group. Astone and Nunez-Wormack (1991) defined minority groups as the population of African-Americans, Hispanics, Asian-Americans, and American Indians that consists of an enormous variety of people from different racial, ethnic, language, and cultural backgrounds. All students from the defined ethnic groups enrolled in technology education or related technology programs are considered minority technology students. Many students are engaged in activities related to technology education but not considered a part of the profession. For example, the national Technology Student Association (TSA) has more than 33,000 elementary students involved in technology-related activities, although many of these school programs are not called technology education. When technology is infused into the regular elementary curriculum, the specific technology activities are not necessarily identified with the description of technology.

DEMOGRAPHICS OF MINORITY STUDENTS

The lack of minority students in technology programs cannot be attributed to the availability of students. Riley (1996), Secretary of Education, predicted that between 1996-2006, total public and private school enrollment will rise from a record 51.7 million to 54.6 million. In the United States, public high school enrollment will increase by 15%; the number of high school graduates will increase 17%, 14% by 2001. Half of the states will have at least a 15% increase in the number of high school graduates, with the Western states having almost a 30% increase in high school graduates.

Hispanic-Americans and Asian-Americans will be the fastest growing segments of the student population. Thus, America's elementary and secondary schools will become more diverse in the next 10 years. Between 1995 and 2005, for example, Hispanic-Americans between the ages of 5 and 17 will increase by 2.4 million. African-Americans in the same age group will increase by 1.1 million. Asian-Americans and other minorities will number an additional 1.1 million. In contrast, the White non-Hispanic student population will grow more slowly, increasing by only 500,000 over the same 10-year span. Currently, African-American and Hispanic youths constitute 27% of the child population. By 2010, these groups will represent nearly 33% of the child population.

According to the National Center for Education Statistics (1995), more than 30% of students in public schools, some 12 million, were from minority groups. Not only are more minority students in public school, but more are completing high school. Major improvements in completion rates were made by African-Americans (from 73% to 86%). However, the completion rate for Hispanic-Americans fluctuated around 70%, with no long-term trend toward improvement. Asian-Americans are completing high school at the highest rate: 91%. Native Americans are completing high school at the lowest rate: 62%. The increase in the population of minority students who have completed high school programs provides a larger pool of potential students for higher education.

Data from the American Association of Colleges for Teacher Education (1994) show an increase in minority student enrollment in higher education over the past five years. In schools, colleges, and departments of education between 1989 to 1991, Hispanic enrollments increased by 44.1%, Native American/Alaskan Natives by 29%, Asian/Pacific Islanders by 22%, and Black/African-American enrollments by 11.9% (American Association of Colleges for Teacher Education, 1994). However, these minorities are not choosing technology education, and the profession should draw upon this valuable resource of minority students.

PROFILES OF STUDENTS IN K-12 TECHNOLOGY EDUCATION PROGRAMS

The literature regarding K-12 programs reveals a gradual increase in the number of minority students in K-12 schools. In contrast, there is a gradual decrease in the number of non-minority students in K-12 schools. This trend indicates the face of public education in the United States is changing. The racial boundaries often defined by the color of one's skin will be more difficult to define because of the increased blending of races and cultures in America.

In K-12 schools, students generally begin their formal exposure to the study of technology in kindergarten classes. Students experience technology in the elementary schools through integrated subject activities. The middle school technology education experience is usually identified as exploratory and is generally required of most students. The K-8 experiences are usually inclusive of all students, regardless of gender or race.

The final stage of the formal K-12 educational experience occurs in the high school. Students in high school programs generally choose to take technology courses they feel will support their career goals. The gender profile of secondary technology students is mostly male from the majority population. However, there are federal and state-funded programs designed to address the lack of females in technology classes. In North Carolina, for example, two Department of Public Instruction-funded Summer Technology Institutes support an increased number of females in technology programs by providing young women in grades 10-12 with high-tech activities and encouraging more females to enroll in secondary technology courses.

Many minority students are enrolled in high school technology programs identified as vocational education programs. Rivera-Batiz (1995) insisted that secondary vocational education programs continue to serve as a "dumping ground" for African-American and other students who are labeled underachievers. Technology education in many states is included as part of vocational education. When the technology program is filled with a majority of underachievers, its overall importance among other school programs is decreased. Students in such programs, many of whom are minority students, are also viewed as not important. Students who feel they are not important generally choose not to participate in the related program activities. Technology education programs defined as high tech or modular or other non-vocational-based programs generally attract a diverse population of students.

The 1990 U.S. Census report shows representation of minority students at all levels of the K-12 formal education program. Of the total K-12 student population, about 30% are minority students. Similarly, selected state supervisors of technology education programs reported that minority students in their respective states comprised 30% of the population of all secondary technology education students. Additionally, the executive director of TSA, whose membership is more than 100,000 students, said that 30% of that population is represented by minority students.

Selected elementary, middle, and high school teachers recognized by ITEA for teaching excellence also reported that between 28% and 30% of their students were minorities. When the K-12 award recipients were asked to profile their minority technology students, they provided similar responses. Minority technology students were no different from other students, they were interested in how things work, they liked hands-on activities, they liked to participate in technology-based and science-related activities, and generally they were just curious about technological devices.

Many minority students participate in student leadership and club activities. For example, at TSA National Conferences, 30% of the students who participate are minorities and are not identified as underachievers. At the 1996 TSA national conference, a selected group of minority students was asked why they participated in TSA events. Their composite responses were that they enjoyed technology, liked the leadership opportunities of TSA, and wanted to continue to participate in technology-related activities in their secondary classes. They said that in their respective schools, Technology Education was the name of the program, and it was not viewed as a program for underachievers.

Minority Students

The National Center for Education Statistics (NCES) (1992) reported that students spend a majority of their time outside of the classroom. How this time is spent may be an indication of the students' growth and opportunities available to them. Participation in extracurricular activities, for example, may affect academic performance, attachment to school, and social development. Almost every high school in the United States offers some extracurricular activities, such as music, academic clubs, and sports. These activities provide opportunities for students to learn the values of teamwork and apply academic skills in other arenas as a part of a well-rounded education; they also provide a channel for reinforcing skills. The NCES report also showed that participation in any extracurricular activity was relatively similar across racial/ethnic groups. However, Blacks were more likely to be involved in school vocational clubs than Whites, Hispanics, and Asians, whereas Asians were more likely than both Blacks and Hispanics to be involved in academic clubs.

Researchers have identified several barriers to student participation in extracurricular activities, ranging from the more tangible, including family or work responsibilities, limited resources for equipment or other expenses, and transportation or other logistical difficulties, to the more complex, such as lack of interest in or alienation from school and its activities (Kleese & D'Onofrio, 1994). For many minority students, extra income is needed to help with family expenses. As a result, many minority students cannot participate in extracurricular events, further reducing the number of minority students in organizations that provide leadership opportunities.

PROFILES OF MINORITY STUDENTS IN POST-SECONDARY PROGRAMS

The profile of postsecondary minority students in technology programs typically mirrors that of minority secondary students, especially at community or technical colleges. Many minority students in two-year colleges took technology-based courses in secondary schools and are continuing their education at the community college. Also, many two-year postsecondary students follow a 2+2 model or Tech Prep model in which they are enrolled in secondary courses and community college courses at the same time (Hirshberg, 1991). In addition, community colleges have always targeted first-generation students (i.e., students who are the first in their families to attend a postsecondary institution and who tend to be from working class families, or to be ethnic minorities, women, or adults) as a primary clientele (Zwerling & London, 1992). However, transfer rates of minority students to four-year colleges and universities are low. Curry (1988) reported that two major problems plague many community colleges. The first problem is high rates of students' attrition, especially among minority students, and the second problem is low rates of successful transfer to four-year colleges.

Minority students' enrollment in four-year colleges and universities increased 27 percentage points between 1972 and 1994 (National Center for Education Statistics, 1997). Minority students who plan to continue their education at a four-year college or university decide early in their high school careers. However, students' plans as high school seniors are likely to reflect their previous academic performance, their financial means, and their educational and career goals. The proportion of all high school seniors in minority groups who planned to continue their education at four-year colleges and universities directly after high school increased between 1972 and 1992, although between-group differences have remained fairly constant.

Postsecondary technology class enrollment is still unequal, with White men as the majority group. According to Flowers (1994), women tend not to participate in technology-based classes, whether in the secondary or postsecondary schools because of the "male-oriented" activities. Some women's perceptions of male-oriented activities in technology classes further reduce the number of women who choose technology education or technology-related careers.

The racial composition of postsecondary students in two-year technology programs shows more African-American students than any other minority group enrolled in community colleges. However, African-American students generally are products of secondary vocational education programs. In addition, many African-American students enrolled in community college programs do not graduate or continue their education at four-year colleges or universities. The small number of minorities who continue their education at four-year colleges and universities is not enough to reduce the major void of minority students in higher education (Curry, 1988). The small number of minorities in higher education continues to exacerbate the problem of minorities who are available to assume leadership roles in the profession of technology education.

Four-year colleges and university technology education programs are experiencing low enrollment of traditional students (students who enter college after high school) and are experiencing more growth in the number of non-traditional students (students who have a degree in another discipline or who enter college after 25 years of age). Like secondary school and community college technology students, students in higher education technology majors are mostly men and non-minority. Current and future trends in population growth and participation in higher education reveal that the number of people of color in the United States is dramatically increasing, but this is a seriously undereducated segment of society. The Association for the Study of Higher Education (1991) found that 27% of all public school students in the 24 largest city school systems are minorities. Yet for nearly all minority groups, high school graduation rates are significantly lower than for the majority, and entry rates of college-age minorities into higher education are actually shrinking.

The Technology Education Collegiate Association (TECA) is the major association for students in technology education programs at the college or university. Unlike the secondary Technology Student Association, the population of this group is less than 15% minorities. Most minority membership comes from historically Black colleges and universities (HBCUs). Also, the gender composition is comprised of less than 15% women. Most women are non-minority.

As illustrated in the profile of the secondary and postsecondary students, participation of minority students continues to decrease. Even in the respective student organizations that foster leadership opportunities, more needs to be done to find ways to motivate more minorities to become involved in the general operations of both associations. TSA and TECA can provide the technology education profession with a source of potential leaders.

OVERCOMING BARRIERS TO INCREASE MINORITY STUDENT PARTICIPATION IN TECHNOLOGY EDUCATION

Many obstacles reduce the number of minorities who participate in technology education. Many problems associated with low minority participation can be found in four major areas, which are in no way inclusive. The four major areas are low recruitment and retention of minorities, the lack of effective mentoring, little or no diversity in programs, and certification issues. In the following sections, each major topic is reviewed, and some solutions to the problems and issues of low minority participation in technology-based courses are discussed.

Recruitment and Retention

The recruitment and retention of all students is a national concern for many college and university programs. Recruitment and retention of minority students is a major goal for most HBCUs. Recent enrollment trends show increases in the number of minority enrollees in colleges and universities.

Attrition is a major problem for American colleges and universities. Efforts to retain students are stymied and made more complex because of the increasing number of enrollees who fit the socioeconomic and demographic profile of "high-risk" students (Jones & Watson, 1990). These high-risk students are defined as minorities, females, low income, and disabled individuals. Although data support an increase in the number of minorities entering college and universities, there is a decline in the number minorities entering the teaching profession (American Association of Colleges for Teacher Education, 1994). To increase minority participation in the technology education profession, more minorities programs.

As the number of minority students in the United States K-12 schools continues to increase, the number of ethnically diverse teachers is declining. By 2000, it is estimated that minority students will comprise 33% of the school population, and that percentage will grow to 39% by 2020 (Johnson, 1991). The number of new recruits to teaching in any subject is insufficient to meet present and projected needs, particularly among minorities. The American Association of Colleges for Teacher Education has predicted that minority teachers will represent fewer than 5% of all K-12 teachers in the U.S. by 2000 (Lankard, 1994).

What are the major reasons for the decreases in the number of minorities in teaching? Webb (1986) attributed the decreases of minorities who enroll in teaching to one major factor. Academically talented minorities now have more career choices available to them than in the past. Other factors that contribute to decreases in minorities in education include low retention rates of minority students attending college and the difficulty of transfer to a four-year institution from a community college (Yopp et al., 1991).

However, before students can enter teacher education programs, they must graduate from high school or receive a General Equivalency Diploma (GED). Jones and Watson (1990) identified several factors associated with attrition and risk of minority secondary students. These factors are grouped into academic, nonacademic, and related factors. Academically, it appears that all students do receive equal preparation in elementary and secondary schools. Moreover, the instructional approaches used by teachers of high-risk students tend to be inappropriate to support different learning styles. Nonacademic factors associated with attrition and risk are generated by both teachers and students. For instance, teachers' negative attitudes might affect students' self-esteem. Thus, many minority students develop low self-esteem and begin to cooperate with systemic forces resulting in pregnancy, dropping out, and delinquency. Related factors are discrimination (e.g., class, gender, race) and differential treatment. Discrimination and differential treatment can undermine students' self-esteem and ultimately facilitate attrition.

Mentoring

Recruitment and retention are important for establishing a pool of minority technology students. However, integral to effective recruitment and retention of minority students is the establishing of an effective mentoring program. Effective mentoring is a supportive relationship between a youth or young adult and someone more senior in age and experience. The older partner offers support, guidance, and concrete assistance as the younger partner goes through a difficult period, enters a new area of experience, takes on an important task, or corrects an earlier problem. During mentoring, mentees identify with their mentors; as a result, they become more able to do for themselves what their mentors have done (Ascher, 1988).

It is important for minority students to have effective mentors available to help them make the transition from their socioeconomic culture to that of the majority students' environment. Typically, a minority student may be the only student in an organization such as the Student Government Association, Science Club, Technology Student Organization, or at the collegiate level, the Technology Education Collegiate Association. Having an effective mentor/mentee relationship will give the minority student immediate access to a person with whom to discuss sensitive matters or other problems associated with being the minority in the group.

Before effective mentorship occurs, the roles of mentors for minority students must be understood, salience and social distance be closed, trust established, and planned realistic expectations for a mentoring program established. Mentors for minority students can range from older, more academically successful students at the next educational stage, to parents, teachers, relatives, or successful people in various careers. The major role of these mentors is to help compensate for inadequate or dysfunctional socialization or give psychological support for new attitudes and behaviors as they simultaneously create opportunities to move successfully in new arenas of education, work, and social life (Ascher, 1988). Consequently, mentors should establish a long-term commitment with their mentee. When the pressures of new life situations, whether in education, work, or social life, have an impact on minority students without effective mentors, they may choose not to engage in additional activities in the new situation.

Another aspect of mentoring is salience and social distance. Effective mentors must explore avenues to close any social distances between the mentor and mentee. Often, the mentor and mentee may not live in the same community. Although it would be more advantageous if the mentor and mentee shared the same community, it is not a necessary requirement for effective mentoring. What is important is the kind of relationship established so the mentee will realize that the mentor is truly concerned about helping him or her close the social gap.

One major method of helping the student understand that the mentor is truly concerned about helping close this social distance is trust. According to Ascher (1988), a critical aspect of any developing mentor-mentee relationship is trust. Ascher further stated that, as a first step, a mentor can build trust by helping the adolescent achieve a very modest goal. The mentor also needs to be personally predictable, and the mentoring program itself should be of some duration. For example, when a secondary technology teacher goes beyond his or her normal duties to establish a Technology Student Association and truly involve minority students, a special bond is created between minority student and that teacher. This bond of trust transcends TSA activities into all aspects of the student's life.

Finally, in establishing effective mentorship, planned mentor activities and realistic expectations must be determined. Mentoring and mentoring programs just do not happen by accident. Professional associations and their states' affiliates should explore the concept of identifying individuals willing to participate in a national mentoring program to help minority students become leaders in technology education.

Planned mentoring programs cannot take a minority student out of a bad situation (e.g., home, school) and immediately transform this student in a new life situation. Mentoring will always be effective only insofar as it accommodates, transforms, or expands the influences of family, school, community, or job. Thus, the power of other influences in the lives of youth must be recognized in any attempt to measure the potential accomplishment of mentoring (Ascher, 1988).

Diversity

Understanding the mind and culture of minority groups requires a special commitment to multicultural understanding and diversity. For teachers at all levels of the educational continuum to become more knowledgeable about multicultural issues, staff training about cultural (and gender) diversity can be a powerful strategy for developing teachers for recruiting and retaining minority students. Skylarz (1993) presented several strategies for enriching teachers' multicultural understanding and suggested several incentives for drawing them to those practices:

1. *Learn a second language.* Teachers will realize a benefit and sense of satisfaction through improved communication within the classroom, other faculty (some of whom will be minority teachers), and parents.

2. *Live in the community.* A better understanding of the school's population is possible when teachers live in the community in which they teach. If living in the community is not possible, the teacher should make several visits to the community in which most students live.

3. *Become involved in the community.* An awareness of community events and participation in some of them can help teachers develop greater understanding of the community's culture(s).

4. *Celebrating cultural events.* With knowledge of the cultural backgrounds of the school population, teachers can work together to organize cultural events or celebrations for the classroom.

A significant element of the recruitment and retention of minority students in secondary and postsecondary programs requires the integration of multicultural activities in the fabric of the total school education program. Skylarz supported this thought by stating that "multicultural understanding will require much more than a plan. It

Certification

For the few minority students who choose technology education as their professional goal, another major obstacle remains in their way: pre-certification qualification. In many states this pre-certification is called Praxis (formally, the National Teacher Exam). Students are required to take this test at the end of their sophomore year or at the beginning of the junior year. In either case, passing the national examination is required for students to be admitted into Schools of Education, teacher education, or student teaching.

If the student passes the pre-certification phase, the final testing at the college or university level is the professional knowledge component. For some disciplines, a specialty area exam is required. Many writers debate the overall validity of these tests, and some writers question whether the tests are racially biased. For example, in states with competency testing, the failure rate for Blacks and other minorities is 2 to 10 times higher than that of Whites (College Board, 1985; Goddison, 1985). As minorities become aware of these statistics, they may reject a teaching career altogether, or at least reject states with competency testing (Hackley, 1985). Recently, in a *Case against a National Test*, dozens of national and local education, civil rights, and advocacy organizations, including those associated with the Fair Test-Initiated Campaign for Genuine Accountability in Education, have consistently opposed any type of national exams at this time. They have argued that all national exam proposals put the cart of testing before the horse of educational reform and that the harmful effects of this effort will fall most heavily on low income and minority-group children (Davey & Neill, 1991).

Overcoming low recruitment and retention of minorities, lack of effective mentoring, little or no diversity in educational programs, and certification issues will indeed be a challenge to the leadership of the technology education profession. Although several strategies have been offered to overcome these barriers, the leadership of the profession must aggressively seek systematically to infuse these strategies throughout all levels of the profession.

SUMMARY

As mentioned earlier, professional associations and affiliates are becoming more proactive in their attempt to raise the consciousness of minority issues such as increased minority participation in leadership roles. Evidence of this raised consciousness is the 1998 Council on Technology Teacher Education yearbook focusing on diversifying the profession. Although, the efforts of the professional associations are well-intended, to achieve equity in leadership roles, more minority students must be recruited and represented at all grade levels. Innovative strategies must be developed and presented to teachers, administrators, parents, and other individuals to reduce the many barriers that keep minority students away from technology-based courses. Early intervention techniques must be put into practice to keep minority students interested in the study of technology and provide motivation to participate in technology activities. Examples of effective organizations that involve minority students in leadership and other participatory competencies are academic clubs and vocational student organizations like the Technology Student Association. Once an effective pipeline is established, the professional associations will have more minority involvement at all levels. The challenge to the leadership of the technology education profession is to establish processes to develop potential leaders from the pool of minority candidates.

America is truly a nation with a diverse population. As demographic data indicate, minorities will become the majority group in the early part of the 21st century. With the minorities becoming the majority, it is the responsibility of the leadership of this nation to prepare minority students for leadership roles. Likewise, our professional associations should do their part in the preparation of minority students to become leaders because "the absence or lack of role models for minority students would result in educational deficits for the nation's youth. Such a deficit would threaten America's future prosperity and ability to compete when compared to other industrialized nations of the world."

REFERENCES

American Association of Colleges for Teacher Education. (1988). *One-third of a nation: A report of the commission on minority participation in education and American life.* Washington, DC: ACE/Education Commission of the States.

American Association of Colleges for Teacher Education. (1994). *Teacher education pipeline III: Schools, colleges, and departments of education enrollments by race, ethnicity, and gender.* Washington, DC: Author.

Ascher, C. (1988). *The mentoring of disadvantaged youth.* ERIC/CUE Digest No. 47.

Association for the Study of Higher Education. (1991). *Pursuing diversity: Recruiting college minority students.* Washington, DC: George Washington University.

Astone, B., & Nunez-Wormack, E. (1991). *Pursuing diversity: Recruiting college minority students.* ERIC Digest. Washington, DC: George Washington University.

College Board. (1985). *Equality and excellence: The education status of Black Americans.* New York: College Entrance Examination Board.

Curry, J. (1988). *The role of the community college in the creation of a multi-ethnic teaching force.* Washington, DC: ERIC Digest.

Davey, L., & Neill, M. (1991). *The case against a national test.* Washington, DC: ERIC/TM Digest.

Flowers, J. (1994). Attention to language: Tips for technology teachers. *The Technology Teacher, 53*(5), 27-30.

Goddison, M. (1985, April). *Testing the basic competencies of teacher education candidates with the pre-professional skills tests (PPSI).* Paper presented at the meeting of the American Education Research Association Annual Conference, Chicago, IL.

Hackley, L. (1985). The decline in the number of Black teachers can be reversed. *Educational Measurement: Issues and Practice, 4*(3), 17-19.

Hirshberg, D. (1991). *The role of the community college in economic and workforce development.* Los Angeles.

Johnson, C. (1991). *Designing strategies for the recruitment and retention of minority students.* Final report. Fayetteville: University of Arkansas.

Jones, D. J., & Watson, B. C. (1990). *"High risk" students and higher education: Future trends.* Washington, DC: George Washington University.

Kleese, E. J., & D'Onofrio, J. A. (1994). *Student activities for students at risk.* Reston, VA: National Association of Secondary School Principals.

Lankard, B. (1994). *Recruitment and retention of minority teachers in vocational education.* ERIC Digest No. 144.

National Center for Education Statistics. (1992).*The condition of education 1995/Indicator 43.* Washington, DC: U.S. Government Printing Office.

National Center for Education Statistics. (1995). *Digest of education statistics table 180.* Washington, DC: U.S. Government Printing Office.

National Center for Education Statistics. (1997). *Findings from the condition of education 1996: Minorities in higher education.* (NCES 97-372). Washington, DC: U.S. Government Printing Office.

Riley, R.W. (1996). *Investing in America's future.* Online document. U.S. Department of Education. http://www.ed.gov

Rivera-Batiz, F. L. (1995). Vocational education, the general equivalency diploma and urban and minority populations. *Education and Urban Society, 27*(3), 313-327.

Skylarz, D. P. (1993). Turning the promise of multicultural education into practice. *School Administrator, 50*(2), 18-20, 22.

U.S. Bureau of the Census. (1990). *Money, income, and poverty status in the United States—1989.* (Current Population Reports, Series P-60, No. 168). Washington, DC: U.S. Government Printing Office.

Webb, M. (1986). *Increasing minority participation in the teaching profession.* ERIC/CUE Digest Number 31.

Yopp, H. K. et al. (1991). The teacher track project: Increasing teacher diversity. *Action in Teacher Education, 13*(2), 36-42.

Zwerling, L. S., & London, H. B. (1992). First generation students: Confronting the culture issues. *New Directions for Community Colleges, 80,* 34.

Section III: INCREASING AND SUPPORTING DIVERSITY

Reading, Writing, and Technology

Karen Coale Tracey
Central Connecticut State University

As the citizens of the United States of America near the end of the century, we acknowledge a world very different from that some perceive of as the "good old days." Often the discussion revolves around the statistics of color and/or gender—"By the year 2000, one out of three Americans will be a member of a racial minority" (Nation, 1986, p. 79), and 52% (100 million) of the total workforce are women (United States Department of Labor, 1993). The participation of women in the workforce has increased since 1947, when women accounted for 31% of the total labor force (Berch, 1982). When we look toward the future job market, minorities and women will be leading the economy.

How can we most effectively educate the majority of our students for the future? In the years ahead, our education system must begin to address those structural changes through which we can empower our students. We must ask why this diversity is so often viewed as a problem to be solved. Why isn't it a wonderful opportunity to build a new, more diverse, more fair, and much more interesting society for all? Well-meaning as it might be, some of the calls to deal with diversity in the classroom sound too much like the Public Health Officer warning us that the chicken pox is coming.

As a democratic nation, the United States has provided a model for other countries to follow. As individuals, we must continually examine how to open the democratic process to all people, regardless of gender or race. The questions that the United States faces are these: How do we create an equitable multiracial/multiethnic society? What role should education play in this process? The task before us is to develop a paradigm for education that is democratic.

The United States of America has always been a nation comprised of a diverse population, beginning with the 500+ language groups of indigenous peoples. Historically, large segments of the

South are home to African-Americans, the Southwest remains predominately Latino, and Native Americans have been concentrated in various areas of the United States. The majority still come from a European background. Statistical data indicate the background of the population in the U.S. is changing from a dominant European background to a population comprised of many other ethnic backgrounds.

Widespread concern about intercultural relations is growing because of the profound changes taking place in the statistical composition of the U.S. population—changes that are causing the U.S. to become an even more culturally diverse nation than ever before. The following highlights are taken from the work of researchers and education writers who have analyzed and commented on these demographic shifts:

- In 1980, five out of six Americans were White; one out of six was Black, Hispanic, or Asian. By 2000, the proportion of Whites will have dropped to two out of three, whereas the minority proportion will have doubled to a third.

- These distinctions mask significant internal diversity. Hispanics, Asians, and immigrant Whites come from many different countries and cultures.

- The White population is both older and less prolific than many other groups. Of the 10 countries sending the most new immigrants to the U.S., 5 are Caribbean, 3 are Asian, and 1 is South American. The only European source of immigrants in the top 10 is the former Soviet Union.

- By 2000, Hispanics will comprise the largest single segment of school-aged children in California and throughout the Southwest. By 2020, California's Whites will account for only 40% of the state's population.

- "Minorities" constitute the majority of public school enrollments in 23 of the nation's largest cities. By 2000 more than 50 major U.S. cities will have a "majority minority" population.

- The school population with limited English proficiency (LEP) has increased by more than 250% in the past decade. Increases in the number of LEP students are occurring even in school districts with declining enrollments. In New York City, 35% of public school students speak a language other than English at home.

(American Jewish Committee, 1989; Banks, 1988; Burstein, 1989; Gay, 1988; Grossman, 1991; Grundy, 1992; Parrenas & Parrenas, 1990)

At the same time that the school-aged population is becoming more multicultural, the teaching profession is becoming more monocultural. In 1985, approximately 88% of the U.S. teaching force was White; by 2000 this is expected to increase to 95% (Burstein, 1989; Pine & Hilliard, 1990; Sleeter, 1989). This imbalance, too, can be a source as well as a result of intercultural tension, because the values and teaching/learning approaches of the predominantly White staff have always worked to the academic and social advantage of White students and to the disadvantage of others (Pine & Hilliard, 1990).

Because learning styles are at least partly determined by culture, comparison of general population and student population demographics is important in planning high-quality educational programming. Therefore, race, class, and gender issues should be discussed in all classrooms (Grant, Sleeter, & Anderson, 1991). To create a suitable learning environment for all students, the teacher must be willing to learn about the students' individual needs, respect students' differences, adapt his or her teaching strategies, and clearly communicate learning expectations (Callahan, 1994).

In this yearbook chapter I will address how curriculum should be delivered to all students, with attention to the learning styles of minority and female students, and the teacher's role in the teaching and learning environment.

DELIVERY OF CURRICULUM

Witkin (1962) first introduced the idea that individuals have different learning styles. A number of scholars have related different learning styles to race, ethnicity, class, and gender. When considering such differences, one must realize that these analyses are generalized, based on social factors, not absolute characteristics of particular groups. Banks (1988) found that different styles of thinking are produced by the experiences of families and groups into which students are socialized. Differences in learning styles can be related to gender as well as race or ethnic group, yet all individuals are not the same and we must not assume that all ethnic groups or all women will have the same learning style. More (1993) explained that

> Most learning styles are learned when students are young children from mother, father, grandparents and close family friends with whom the child interacts regularly. From them, the child learns content and skills. But the child also "learns how to learn." The learning styles of the caregivers have considerable influence on the child's learning styles. By the time a child gets to school, many of the learning styles have already been laid down. (pp. 8-9)

If traditional teaching practices and classroom structures have been designed by and for a White male culture, female and minority students may find it harder to learn in those environments. Helman (1990) defined cultures as "a set of guidelines (both explicit and implicit) which individuals inherit as members of a particular society, and which tells them how to view the world, how to experience it emotionally, and how to behave in relations to other people and to the natural environment" (pp. 2-3). Anderson (1989) contended that preference for learning environments is rooted in the cultural backgrounds of the students. Traditional students, mostly from European cultures, have a different preference for learning environments than their counterparts who have other ethnic backgrounds. Men from a European culture developed most institutions and organizations; hence, a cultural learning style is established for that group of learners and/or workers. A cultural learning style is the way groups of people within a society or culture learn, pass on, interpret, and use new information. Learning styles are influenced by how a culture socializes its children and young people. Richard Paul (1990) suggested several strategies to be used in the teaching of all students. These pedagogical strategies are appropriate to meet the needs of all learners in any classroom setting.

Sandhu (1994) summarized Anderson's classification of students in "Cultural Diversity in Classrooms: What Teachers Need to Know." Anderson (1989) categorized our society into two broad learning groups: traditional—European heritage and diverse—other ethnic groups. Anderson described the traditional student's learning environment this way: "Competitive learning environment; individual study; increase distance when communicating; communicating style is more formal and rigid; express emotions selectively; task completion takes primacy; and process relevant or irrelevant information efficiently." Anderson also described the learning environment for the diverse student: "Cooperative learning environments; group study; minimal distance when communicating; communication style is informal and conversational; express emotions freely; task orientation relational to personal demands; and seeks personal relevance when processing information" (Sandhu, 1994, p. 8). Anderson's model is quite different from the more traditional five dimensions of learning styles model: Global/Analytic, Verbal/Imaginable, Concrete/Abstract, TEF/Reflective, and Modality. Generally, an individual does not "fit" all aspects of one model and therefore should not be looked at in a narrow frame. It is very important to focus on the most effective model for the students to learn and excel.

Once the teacher understands that all students learn differently, then the classroom will become an equal playing ground for all. Okebukola (1986) suggested that the learner's cultural background could have a greater effect on education than does the substantive nature of the course content. Furthermore, unless students can apply what is being taught to themselves, especially in terms of their cultural background, many teaching strategies used by teachers are likely to be less than effective in enhancing learning.

CULTURAL LEARNING STYLES

The National Science Teachers Association (NSTA) (1991) policy statement on multiculturalism lists learning style as an important concern for science teachers. Historically, science educators have called for science teachers to use learning style information to improve instruction (Bonnstetter, Horne, & McDonald, 1991; Kuerbis, 1988). Claxton and Murrell (1987) stated that the most important need in learning-style research is to identify the learning style of children of color. Teachers must be aware of the various cultures in the classroom and actively seek knowledge about cultures by reading and contacting regional/community resource people. Teachers must accept and treasure students' culture and values; teachers must also empower the students through personal attention, encouragement, and support (Sandhu, 1994). It is important to understand the students' cultures for all students to be active participants in the classroom.

African-American Learning Styles

Hale (1986) described the learning styles of African-American children in her book, *Black Children: Their Roots, Culture, and Learning Styles*. According to Hale, most young African-American children are perceived as successful in their homes, churches, and communities. A failure pattern in some children is evident only after a few years in a school designed by and for European-American or Western values that are in some ways alien to the African-American way of life. Sandhu (1994) described the most conducive learning environment for African-American children is a bright classroom. The teacher should give specific directions/explanations to the students prior to starting or completing a task. The students feel more comfortable when someone with authority or special knowledge is present. They learn best when initially listening to a verbal instruction such as a lecture, discussion, or routine. African-American students are highly tactile and feel a strong need to keep their hands busy when they are thinking hard. Because they are highly kinesthetic, they require whole body movement and/or real life experiences to absorb and retain material to be learned. Students want to achieve to please their parents or parent figure. Last, Hale (1986) suggested that African-American children hold values and personal belief systems as more important than logic and abstractions. These children like working cooperatively rather than competitively, and the African-American culture promotes the community above the individual. Therefore, teamwork and cooperative learning can be an effective method of teaching at all grade levels.

Latin/Hispanic American Learning Styles

The deepest belief for Latinos is the preeminence of family and community. This belief is manifested by a need to protect family members and to be self-sacrificing (Fitzpatrick, 1987). Griggs and Dunn (1996) described a greater inclination among Hispanic than Anglo adolescents to adopt their parents' religious beliefs and lifestyle; Hispanic male adolescents seek independence earlier than male adolescents of the general U.S. population. Several researchers have compared students of various ethnic groups in terms of five categories of learning style. These studies suggest that cool temperatures and formal design of an environmental learning style are important for Mexican-American students. Latinos' strongest perceptual strength is a kinesthetic physiological learning style, and Mexican-American students are more field dependent than are non-minority students, who generally prefer a psychological learning style.

European American Learning Styles

The literature describes Anglo students as preferring to study with peers through discussions and interactions. They may easily learn alone and easily recall what has been read or observed. Anglo students need variety as opposed to routines, and their primary perceptual strength is visual. They often eat, drink, chew, or bite objects while concentrating, and they prefer evening as the best time for study (Sandhu, 1994). Anglo students may possess one or more learning styles. These characteristics many times do not reflect the home environment but are rather a learned behavior from being in the school system. The current structure for education was created at the turn of the century, when scientific management and proscribed procedures were the order of the day. Although these factory-model structures served the public well for decades, today's society requires new ways of reaching students. This is especially true for students who often "fall through the cracks" in a factory model (Bess, 1996).

Many people are surprised to think of schools as factories; however, their very structure mirrors the industrial model of production developed by European Americans. Dr. Gary Watts, former head of the National Education Association's National Center for Innovation, described the traditional high school as an assembly line. Students typically get 45 minutes of instruction in a discrete, isolated subject. It doesn't matter that some students can learn the information or skill in 10 minutes and others need a day and a half. All students will receive the same 45 minutes of instruction each day for the school year (Newberry, 1996). This ongoing routine ignores the reality of different learning speeds and styles.

Native American Learning Styles

A Native American child comes from a culture in which children learn mainly by listening and not interrupting. These children are more likely to develop a more reflective learning style. If a child comes from an environment in which learning is demonstrated or symbols and images are used regularly, then the child is more likely to develop a more imaginable learning style. If a child comes from a culture in which the spoken or written word is used a great deal in learning (as in Western, middle class cultures), that child will be more likely to develop a more verbal learning style (More, 1993). Native American students in many situations have a tendency to be more skilled in performing tasks than in verbal expression, they are more visual than auditory linguistic and more oriented toward observation or imitation than toward verbal instruction, and they are more comfortable with spatial than with sequential activities and with group, peer, or cross-age learning projects than with individual question-and-answer sessions (Moore, 1988).

Learning Styles of Female Students

Gender differences are mostly found in four areas of psychological studies: aggression, spatial ability, verbal ability, and mathematical ability. Maccoby (1980) found that males are overwhelmingly more aggressive than females. In tasks of spatial ability, Linn and Peterson (1986) indicated that males have more ability than females visually to rotate objects in space. The gender differences in spatial performance tend to appear reliably by age 10 (Johnson & Meade, 1987). This gender imbalance can be closed or corrected by introducing all young children to a variety of toys as children and encourage the girls to build and analyze as boys have been encouraged for generations. In verbal ability, females tend to have a slightly higher average than males in verbal ability tasks such as spelling and comprehension (Finucci & Childs, 1981), which does not become evident until primary school (Linn & Peterson, 1986). In mathematical ability tasks, males tend to obtain higher scores than females (Benbow & Stanley, 1983; Randhawa & Hunt, 1987).

Sharps, Welton, and Price (1993) found that the gender of the individual does not limit the possible range of cognitive abilities. Their study indicated that gender might not be an indicator in the ability to view objects spatially. The perceived gender difference may result from the placement of individuals in an environment in which they do not have the opportunity to try various tasks (Tracey, 1995). Females are expected to do as poorly in some subjects as members of certain racial and ethnic groups are expected to do (Schwartz & Hanson, 1992). Research over the last decade has shown that males and females have different classroom experiences because they approach learning differently and teachers have a tendency to treat male and female students differently.

Evidence exists that males and females tend to approach learning from different perspectives, although the reasons for differences continue to be debated. Many societies explicitly define the proper gender roles of family and professional life. From infancy, individuals are faced with how they are to conduct themselves in terms of what is proper for family and professional involvement in a social setting. Therefore, society's culture maintains a set of boundaries that one is expected to recognize and a set of norms to follow (Gallein, 1992).

In the classroom, females prefer to use a conversational style that fosters group consensus and builds ideas on top of each other. Males learn through argument and individual activity—behaviors fostered early. Most classroom discourse is organized to accommodate male learning patterns (Ong, 1981). It is important for schools to go beyond equal access in attempting to balance differences in exposure. Teachers should provide opportunities for females to be more involved with and persist in using computers and other technological devices. Females should be educated and encouraged from the age of five that mathematics, science, and technology are important and relevant to their lives. Mathematics, science, and technology teachers as well as guidance counselors can play a role in suggesting education and employment opportunities in technology-related fields.

FACTORS INFLUENCING PARTICIPATION IN TECHNOLOGY EDUCATION

Silverman and Pritchard (1993) identified three barriers to greater participation of girls in technology education: lack of information regarding technology education and careers related to applied technology, lack of connection to technology and the ways that their coursework applies to their future, and lack of flexibility in school schedules. Silverman and Pritchard (1993a), in their study entitled "Building Their Future: Girls in Technology Education in Connecticut," found the following:

1. In middle school, girls appear to enjoy technology education and have confidence in their abilities, but emerging sexism among peers begins to differentially affect participation on the basis of gender. Girls may also respond more positively to some projects and be more interested in some aspects of their technology education classes that include gender neutral or traditionally female-identified activities. (p. 3)

2. Girls are discouraged from taking more technology education in high school because stereotypes about appropriate careers

for women are still segregating, girls don't know enough about technological careers, don't connect what they are learning in the classroom with careers, and are uninformed about economic realities and the world of work. (p. 7)

3. The high school survey suggests that although girls who take technology education in high school are willing to challenge stereotypes about technology as a male occupation, they have less confidence in their abilities and are thinking less in terms of well-paid jobs than the boys in their classes. (p. 12)

The educational system must address how it is failing to support specific groups of students and how to remedy the situation. Administrators, teachers, guidance counselors, and parents must all work together to create an environment that best supports learning.

THE ROLE OF TEACHERS IN LEARNING

Teachers' intercultural knowledge, attitudes, and behaviors have powerful effects on the quality of intercultural relations in schools and classrooms. Therefore, it is important to consider the research on practices designed to prepare teachers successfully to teach culturally diverse student populations and promote harmonious relations among them. Larke (1990) wrote

> Studies have shown that a high correlation exists among educators' sensitivity (attitudes, beliefs and behaviors toward students of other cultures), knowledge and application of cultural awareness information, and minority students' successful academic performance. Effective teachers in diverse settings have been found to exhibit high levels of cultural sensitivity (which is) exhibited by the modified curriculum and instructional designs they incorporate to ensure that all students achieve excellence and equity. (p. 24)

Researchers have found that Anglo females and African-American, Native American, and Hispanic-American females and males fall into the relational field-sensitive end of the learning style continuum. Anglo males and Asian-American males have a tendency to be more analytical field independent. These differences have an impact on the way a teacher delivers material in the classroom and how well a student interprets and understands the material (Anderson & Adams, 1992). The following matrix compares the relational and analytical learning styles.

Teachers must be aware of the various cultures that exist in the classroom. They must actively seek knowledge about cultures by reading and by contacting regional/community resource people. Teachers must accept students' culture and values and look on them as a treasure. Teachers must also empower students through personal attention, encouragement, and support (Sandhu, 1994).

Table 1. Characteristics of Students' Learning Styles

Relational Style	Analytical Style
Perceive information as part of total picture	Able to dissemble information from total picture (focus on detail)
Exhibit improvisational and intuitive thinking	Exhibit sequential and structured thinking
More easily learn material that have a human, social content and characterized by experiential/cultural relevance	More easily learn materials that are inanimate and impersonal
Good memory for verbally presented ideas and information, especially if relevant	Good memory for abstract ideas and irrelevant information
More task-oriented concerning nonacademic areas	More task-oriented concerning academics
Influenced by authority figures' expression of confidence or doubt in students' ability	Not greatly affected by the opinions of others
Prefer to withdraw from unstimulating task performance	Show ability to persist at unstimulating tasks
Style conflicts with traditional school environment	Style matches most school environments

(Anderson & Adams, 1992, p. 23)

What can we learn from the research on efforts to promote accurate knowledge, positive attitudes, and effective instructional and social behaviors on the part of teachers? For one thing, research shows that brief and superficial training may increase knowledge but has little or no effect on attitudes or behavior (Bennett & Jay, 1997; Grottkau & Nickolai-Mays, 1989; Larke, 1990; Merrick, 1988; Sleeter, 1989).

As with the research involving children and youth, the research on preservice and inservice educators indicates that "one-shot" overviews or short-term courses do not produce the outlook and skills needed to work successfully with diverse populations. Larke (1990) wrote

> It seems very important that teacher educators recognize and respond to this concern by providing more than one isolated course. One course is insufficient to change the attitudes and behaviors of preservice teachers to appreciate, accept, and respect the diversity of students facing them in future classrooms. (p. 29)

In a classroom, teachers set the standard for learning. In relying on traditional teaching methods, they adhere to traditional norms and beliefs about cultural gender differences, which primarily benefit male students, who make up the majority (Kramaerae & Treichler, 1990; Sandler, 1982). Silverman and Pritchard (1993b) recommended that teachers have the opportunity to discuss equity issues through a number of different forums, including workshops with outside facilitators and in-school meetings to discuss guidelines. The forums should involve training to understand the differing needs and interests of all students, including group versus individual projects, behavior in the classroom among boys and girls that ensures that girls play an equal role, and ways to make the classroom and subject matter more attractive to girls, including choices about projects, whether design and decoration can be given credit on a project along with mechanical aspects.

Researchers have also identified numerous teacher behaviors and other classroom variables as being associated with positive intercultural environments (Aviram, 1987; Campbell & Farrell, 1985; Garcia, Powell, & Sanchez, 1990; Sanders & Wiseman, 1990). Not surprisingly, many of these features are commonly found in connection with positive classroom environments in general. According to the previously listed researchers, teachers in interculturally harmonious environments focus on the climate and tone of the learning environment by

- Arranging their classrooms for movement and active learning
- Interacting one-to-one with each child at least once daily
- Communicating high expectations for the performance of all students
- Giving praise and encouragement to all students
- Communicating affection for and closeness with students through verbal and nonverbal means such as humor, soliciting student opinion, self-disclosure, eye contact, close proximity, smiling, and so on
- Avoiding public charting of achievement data
- Giving children responsibility for taking care of materials, decorating, greeting visitors, etc.
- Treating all students equally and fairly

Classrooms should also reflect the ethnic heritage and background of all children in the classroom. Teachers can create a multicultural environment by

- Displaying artifacts (weavings, photos, etc.) from various cultures, as found in textbooks, information books, magazines, journals, and other instructional materials
- Incorporating the displayed items in learning activities
- Changing items and topics throughout the year

INCREASING A POSITIVE IMAGE THROUGH LANGUAGE

A positive self-concept is necessary for students to develop a sense of well-being, to be motivated in the classroom, and to be successful in both the classroom and private life. An individual's self-concept can be affected by language, whether spoken or found in written assignments and/or textbooks. A new vocabulary is needed to motivate all students in a classroom. Donna Koppi Boben's *Guidelines for Equity Issues in Technology Education* (1985) gives many suggestions and examples to assist in developing a gender-neutral vocabulary and developing roles and traits in the classroom. Her guidelines for developing language are these:

- Use gender-fair language throughout written materials in order to successfully portray equitable options for both genders.
- Refer to men and women in a consistent manner.
- Use gender-fair or gender-neutral language.
- References to jobs, roles, and personal characteristics should be gender free.
- Terms or adjectives that patronize or trivialize women and girls should not be used.
- Use gender-free words instead of gender-specific words to describe occupations or positions.
- Use words and phrases that are inclusive of both genders.
- Plural pronouns are acceptable substitutes for the masculine singular.
- Editorial policies should require the use of inclusive terms for the alternating gender referents.
- Parallel language should be used when referring to or describing women and men.
- Women and men should be referred to by name in the same style and manner.

- Use the words *woman, female,* or *women* for adult women in the same way as the words *man, male,* or *men* are used for adult men. (pp. 6-8)

Language does not stop at the gender line; it is a part of all cultures. Language reflects a society's attitudes and thinking about different cultures and religions. White Anglo-Saxons have been the dominant group in most parts of the United States since the 18th century. This race has not been subjected to mainstream verbal abuse represented by the English language that people of other cultures have received. It is important for educators to recognize racism in language and to increase their awareness of terminology that develops a positive self-image. Listed below are guidelines to assist in developing appropriate language for a classroom:

- Check with organizations in your area to verify acceptable terms for racial groups in a particular state or region.
- Use the currently preferred terms for the major racial groups in your area. In most areas, the following terms are acceptable: African-American, Native American, Asian-American, Latino, and European American (Anglo).
- Do not use racial identification. Descriptions of individuals or groups of people should be relevant to the context of the statement, and not used to qualify a racial descriptor.
- Test for the use of qualifiers by substituting the word *White* in place of *Hispanic,* or other terms describing the race of a person. Also, substitute an Anglo surname for a Hispanic or an Asian one to test for qualifiers. If the substituted qualifying word proves to be irrelevant to the statement, drop the qualifier from the statement.
- Describe or use examples of people of all races equally in rural settings, suburban areas, and urban areas.
- Depict equal numbers of people of all races among the well-to-do, the middle, and poor classes of people.
- Review descriptive words in written and spoken communication for implications of bias and stereotypes.

- Replace biased and stereotypic words with words that have positive connotations for the person or group they describe.
- If it is not relevant or appropriate, do not attach adjectives to describe people of any race. (Boben, 1985, pp. 12-15)

Bringing students into task-related and social contact with those who are culturally different can provide an opportunity to change the attitudes of students for the better.

DIVERSE TEACHING STRATEGIES FOR DIVERSE LEARNERS

Researchers have found that the use of cooperative learning as a mechanism to improving intercultural relationships is supported by more well-designed research than any other single schooling practice. Organizing learners into culturally heterogeneous teams, giving them tasks requiring group cooperation and interdependence, and structuring activities so that teams can experience success comprise an extremely powerful means of enhancing intergroup relations (Parrenas & Parrenas, 1990; Slavin, 1990; Swadener, 1988; Warring, Johnson, Maruyama, & Johnson, 1985). In small groups, students are expected to work together and share ideas. These learning groups have been shown to enhance self-esteem (Callahan, 1994). Thus, students should use current techniques in cooperative learning groups to make cultural and technological transitions in the classroom.

Cooperative learning also honors the preferred learning styles of African-Americans, Latinos, Native Americans, and Anglo female students. In fact, cooperative learning holds true for learners of all ages, including adults, for virtually every kind of difference—race, ethnicity, handicap, gender, academic ability, and so on.

Individuals who participate in cooperative learning activities typically show an increase in the number of cross-cultural contacts and friendships they develop and value through these activities. Research also indicates a decrease in intergroup tension by those who participated in cooperative learning activities and those who have observed the activities. Pate (1988) described the cooperative learning environment as one that fosters the need for individuals to share common problems, tasks, goals, and success with people from another ethnic group. In this environment, people develop positive feelings toward them.

Cotton (1993) summarized the positive intergroup outcomes from cooperative learning activities. They include and are not limited to

- Positive interdependence. Students perceive that they need each other to complete the group's task and that they will "sink or swim together."
- Individual accountability. Each student's performance is frequently assessed, and the results are given to the group and to the individual.
- Group processing. Groups discuss how well they are achieving their goals and maintaining effective working relationships among members. Teachers give feedback on how well the groups are working together and input for improvement.
- Interpersonal and small-group skills. Teachers teach social and process skills needed for effective group functioning, including leadership, decision making, trust building, communication, and conflict management.
- Face-to-face promotive interaction. Students promote each other's learning by helping, sharing, and encouraging one another's efforts to learn. Students explain, discuss, and teach what they know to their groupmates.

Teachers can use computers and software to support a variety of learning styles representative among various cultural groups in a classroom. Children can work alone or with others, depending on personal preference and the nature of the task. In this way the teacher can support field sensitivity and field independence. Children can explore computer capabilities through self-discovery or learn the step-by-step program procedures from the teacher and their peers. Holistic and analytical learners benefit from the flexibility computers can offer. Many software applications provide animated graphics, sound input and output, and visual text for visual and auditory learners. Keyboarding and use of the mouse assist kinesthetic learners, as does physical mobility, which allows students to move to different areas of the room. Giving students choices in how they learn best allows them the opportunity to engage in decision making and to select activities that meet their needs and interests (Chisholm, 1996; National Council of Teachers of Mathematics, 1989). When access to computers is equitable, technology improves opportunities for diverse student populations. Computer-based labs and data analysis allow students to connect mathematics and science to real issues in their communities (Sheppo, Hartsfield, Ruff, Jones, & Holinga, 1994; Smith, 1989).

Strategies that allow teachers to spend extra time with individual students are especially important (Callahan, 1994). The technology laboratory allows teachers to supervise students, diagnose the students' progress, and assist students on an individual basis. Cultural factors may deter students from appropriate participation in classroom and laboratory activities. A willingness to learn about the learning styles of each student is important to student success. Educators must respect differences in learners and be willing to adapt their strategies better to communicate their expectations (Callahan, 1994).

SUMMARY

Caruthers (1996) encouraged teachers to become facilitators of the learning process and students to take responsibility for their learning and to explore their world. To accomplish these goals, Caruthers recommended educators must

- Understand the impact of prejudice, bias, and stereotyping
- Receive ongoing feedback about individual behaviors and practices and ways to eliminate behaviors that are damaging to students
- Give attention to acquiring knowledge of human development and learning, including cultural norms and traditions, learner-centered practices, cultural socialization, and learning styles
- Place more emphasis on language development provided through teacher-student dialogue and the experiences of

children, including understanding second language acquisition, and the socio-cultural aspects of language
- Use the abilities, skills, talents, and strengths of all students to expand and extend their learning and achievement, giving attention to questioning strategies, higher order thinking, and the application of knowledge
- Capitalize on the social skills that children typically bring to the classroom and organize classrooms around a more active, participatory approach
- Acquire strategies for improving the participation of minorities and girls in the classroom, providing more academic attention, developing a system for calling on them more often, and encouraging minorities and girls in math and science. (p. 7)

The history of technological development and the use of positive role models can aid in teaching technological concepts in a multicultural classroom. For example, teachers can introduce some renowned technologists and inventors belonging to the various cultural groups within the class. In the learning process, students need to come to see that only they can learn for themselves and that they will not do so unless they actively and willingly engage themselves in the process. Teachers should, however, provide opportunities for students to decide what they know and help them develop strategies for finding or figuring it out (Paul, 1990). Students should also keep journals, develop portfolios for their work, and set questions to guide future discussions and studies to help facilitate the learning process (Bagley & Gallenberger, 1992).

REFERENCES

American Jewish Committee. (1989). *Educating the newest Americans: Report of the task force on new immigrants and American education.* New York: American Jewish Committee/Columbia University Institute for Urban and Minority Education.

Anderson, J. A. (1989). Reaching beyond monoculturism: Acknowledging the demands of diversity. *SAEOPP Journal, 7,* 4-13.

Anderson, J., & Adams, M. (1992, Spring). *Teaching for diversity.* (Eds. L. Border & N. Chism). San Francisco: Jossey-Bass.

Aviram, O. (1987). The impact of school as a social system on the formation of student intergroup attitudes and behavior. *Journal of Educational Equity and Leadership, 7*(2), 92-108.

Bagley, T., & Gallenberger, C. (1992). Assessing students' disposition: Using journals to improve students' performance. *The Mathematics Teacher, 50*(8), 660-662.

Banks, J. A. (1988). Cultural diversity and intergroup relations: Implications for educational reform. In C. A. Heid (Ed.), *Multicultural education: Knowledge and perceptions.* Bloomington: Center for Urban and Multicultural Education, Indiana University.

Benbow, C. P., & Stanley, J. C. (1983). Sex differences in mathematical ability: More facts. *Science, 222,* 1029-1231.

Bennett, L., & Jay, J. (1997). *Elementary teacher preparation with multicultural requirements.* Missouri, ED405294.

Berch, B. (1982). *The endless day: The political economy of women and work.* New York: Harcourt Brace Jovanovich.

Bess, S. (1996). *Overlooked students.* Discovery Communications, Inc.(On-line) http://school.discovery.com/vvault/ttv/archive/ttvep37.html.

Boben, D. K. (1985). *Guidelines for equity issues in technology education.* Reston, VA: International Association of Technology Education.

Bonnstetter, R., Horne, S., & McDonald, D. (1991). On reach: Use a variety of styles to meet the needs of everyone in your class. *Science Scope,* 15(3), 48-49.

Burstein, N. D. (1989). Preparing teachers to work with culturally diverse students: A teacher education model. *Journal of Teacher Education,* 40(5), 9-16.

Callahan, W. (1994). Teaching middle school students with diverse cultural backgrounds. *The Mathematics Teacher,* 87(2), 122-126.

Campbell, R. L., & Farrell, R. V. (1985). The identification of competencies for multi-cultural teacher education. *Negro Educational Review,* 36(3-4), 137-144.

Caruthers, L. (1996). *Classroom interactions and achievement. What's noteworthy on learners, learning & schools.* Mid-continent Regional Educational Laboratory (McREL). [On-line]. http://www.mcrel.org. http://www.mcrel.org.

Chisholm, I. M. (1996). *Computer use in an elementary multicultural classroom.* [On-line]. http://www.coe.uh.edu/insite/elec_pub/html 1995/033.htm.

Claxton, C. S., & Murrell, P. H. (1987, September). Experiential learning theory as a guide for effective teaching. *Counselor Education and Supervision,* 27, 4-14.

Cotton, K. (1993, November). Topical synthesis #7 fostering intercultural harmony in schools: Research finding. *School Improvement Research Series* (SIRS). [On-line]. http://www.nwrel.org/scpd/sirs/8/topsyn7.html.

Finucci, J. M., & Childs, B. (1981). Are there really more dyslexic boys than girls? In A. Ansara, N. Geschwind, A. Galaburda, M. Albert, & N. Gartrell (Eds.), *Sex differences in dyslexia.* Towson, MD: The Orton Dyslexia Society.

Fitzpatrick, J. (1987). *Puerto Rican Americans.* Englewood Cliffs, NJ: Prentice Hall.

Gallein, A. A. (1992). *Selected factors related to career achievement of women administrators in community colleges.* Unpublished doctoral dissertation, North Carolina State University.

Garcia, J., Powell, R., & Sanchez, T. (1990, April). *Multicultural textbooks: How to use them more effectively in the classroom.* Paper presented at the annual meeting of the American Educational Research Association, Boston.

Gay, G. (1988). Designing relevant curricula for diverse learners. *Education and Urban Society, 20*(4), 327-340.

Grant, C. A., Sleeter, C. E., & Anderson, J. E. (1991). The literature on multicultural education: Review and analysis. *Educational Studies, 12*(1), 47-71.

Griggs, S., & Dunn, R. (1996). *Hispanic-American students and learning style.* Urbana, IL: Clearinghouse on Elementary and Early Childhood Education.

Grossman, H. (1991). Multicultural classroom management. *Contemporary Education, 62*(3), 161-166.

Grottkau, B. J., & Nickolai-Mays, S. (1989). An empirical analysis of a multicultural education paradigm for preservice teachers. *Educational Research Quarterly, 13*(4), 27-33.

Grundy, T. (1992, December). ESL/Bilingual education: Policies, programs, and pedagogy. *OSSC Bulletin, 36*(4).

Hale, J. (1986). *Black children: Their roots, culture and learning styles.* Baltimore, MD: Johns Hopkins University Press.

Helman, C. G. (1990). *Culture, health, and illness* (2nd ed.). Boston: Wright.

Johnson, E. S., & Meade, A. C. (1987). Developmental patterns of spatial ability: An early sex difference. *Child Development, 58,* 725-740.

Kramaerae, C., & Treichler, P.A. (1990). Power relationships in the classroom. In S. L. Gabriel & I. Smithson (Eds.), *Gender in the classroom: Power and pedagogy* (pp. 41-59). Chicago: University of Illinois Press.

Kuerbis, P. (1988, March). Research matters . . . to the science teacher: Learning styles and science teaching. *NARST, 3*(1).

Larke, P. J. (1990). Cultural diversity awareness inventory: Assessing the sensitivity of preservice teachers. *Action in Teacher Education, 12*(3), 23-30.

Linn, M. C., & Peterson, A. C. (1986). Meta-analyses of gender differences in spatial ability. In J. Hyde & M. Linn (Eds.), *The psychology of gender: Advances through meta-analysis.* Baltimore, MD: Johns Hopkins University Press.

Maccoby, E. (1980). *Social development: Psychological growth and the parent-child relationship.* New York: Harcourt Brace Jovanovich.

Merrick, R. M. (1988). *Multicultural education: A step toward pluralism.* Indiana, ED302451.

Moore, C. (1988). The implication of string figures for American Indian mathematics education. *Journal of American Indian Education,* 28(1), 16-25.

More, A. J. (1993). *Adapting teaching to the learning styles of Native American students.* British Columbia, Canada.

A nation prepared: Teachers for the 21st century. (1986). *The Report of the Task Force on Teaching as a Profession.* New York: Carnegie Forum of Education and the Economy.

National Council of Teachers of Mathematics (NCTM). (1989). *Curriculum and evaluation standards for school mathematics.* Reston, VA: NCTM.

National Science Teachers Association. (1991, October/November). NSTA releases position paper on multicultural science education. *NSTA Reports,* p.1.

Newberry, G. (1996). *The outmoded factory model.* Discovery Communications, Inc. [On-line]. http://school.discovery.com/vvault/ttv/archive/ttvep37.html.

Okebukola, P. A. (1986). The problem of large classes in science: An experiment in cooperative learning. *European Journal of Science Education,* 8(1), 73-77.

Ong, W. (1981). *Fighting for life.* Ithaca, NY: Cornell University Press.

Parrenas, F. Y., & Parrenas, C. S. (1990). *Cooperative learning, multicultural functioning, and student achievement.* San Bernardino, CA: San Bernardino School District.

Pate, G. S. (1988). Research on reducing prejudice. *Social Education,* 52(4), 287-289.

Paul, R. (1990). *Critical thinking: What every person needs to survive in a rapidly changing world.* California: Sonoma State University, Center for Critical Thinking and Moral Critique.

Pine, G. J., & Hilliard, A. G. (1990). Rx for racism: Imperatives for America's schools. *Phi Delta Kappan, 71*(8), 593-600.

Randhawa, B. S., & Hunt, D. (1987). Sex and rural-urban differences in standardized achievement scores and mathematics sub-skills. *Canadian Journal of Education, 12,* 137-151.

Sanders, J. A., & Wiseman, R. L. (1990). The effects of verbal and nonverbal teacher immediacy on perceived cognitive, affective, and behavioral learning in the multicultural classroom. *Communication Education, 39*(4), 341-353.

Sandhu, D. S. (1994). *Culturally specific learning styles: Some suggestions for teachers.* ED 370 910.

Sandler, B. (1982). *The classroom climate: A chilly one for women?* Washington, DC: Association of American Colleges.

Schwartz, W., & Hanson, K. (1992). *Equal mathematics education for female students.* Washington, DC: Office of Educational Research and Improvement.

Sharps, M. J., Welton, A. L., & Price, J. L. (1993). Gender and task in the determination of spatial cognitive performance. *Psychology of Women Quarterly, 17,* 71-83.

Sheppo, K. G., Hartsfield, S. J., Ruff, S., Jones, C. A., & Holinga, M. (1994). How an urban school promotes inclusion. *Educational Leadership, 52*(4), 82-84.

Silverman, S., & Pritchard, A. (1993a, September). *Building their future: Girls in technology education in Connecticut.* Hartford: Vocational Equity Research, Training and Evaluation Center, Connecticut State Dept. of Education.

Silverman, S., & Pritchard, A. (1993b, September). *Guidance gender equity and technology education. vocational equity research, training and evaluation center.* Hartford: Vocational Equity Research, Training and Evaluation Center, Connecticut State Dept. of Education.

Slavin, R. E. (1990). Research on cooperative learning: Consensus and controversy. *Educational Leadership, 47*(4), 52-54.

Sleeter, C. E. (1989). Doing multicultural education across the grade levels and subject areas: A case study of Wisconsin. *Teaching & Teacher Education, 5*(3), 189-203.

Smith, L. B. (1989). *A catalog of successful math programs across Alabama, Florida, Georgia, Mississippi, North Carolina, and South Carolina.* (Vol. II). Research Triangle Park, NC: Southeast Educational Improvement Laboratory.

Swadener, E. B. (1988, April). *Teaching toward peace and social responsibility in the early elementary years: A friends school case study.* Paper presented at the annual meeting of the American Educational Research Association, New Orleans, LA.

Tracey, K. C. (1995). *Expressed attitudes of university administrators and faculty members toward women teaching in Bachelor degree granting industrial technology programs.* Unpublished doctoral dissertation, University of Massachusetts-Amherst.

United States Department of Labor, Bureau of Labor Statistics. (1993). *Geographic profile of employment and unemployment.* Washington, DC: U.S. Government Printing Office.

Warring, D., Johnson, D. W., Maruyama, G., & Johnson, R. (1985). Impact of different types of cooperative learning on cross-ethnic and cross-sex relationships. *Journal of Educational Psychology, 77*(1), 53-59.

Witkin, H. A. (1962). *Psychological differentiation.* New York: John Wiley.

Mentors for Women in Technology

Daniel L. Householder

Texas A & M University

Women who pursue careers in technology face challenges that are substantially different from those faced by their male colleagues. In this chapter, I explore the literature on mentoring women as they complete their graduate education and pursue their careers in technology. Guidelines for effective mentoring are reviewed, and benefits to the mentors and the protégées are explored. The chapter is concluded with a discussion of the benefits that accrue to the profession by effectively mentoring women in technology.

IMPORTANCE OF MENTORING DURING GRADUATE STUDY

Women's need for mentoring support is most apparent during the graduate school experience. However, the Committee on Women in Science and Engineering (1991) pointed out that graduate education increasingly serves the needs of the (predominantly male) faculty and the institution rather than the needs of graduate students. In addition, women are less likely than men to have teaching or research assistantships and the concomitant opportunities for interaction with peers and faculty.

During graduate study, the mentor helps the protégée by coaching, teaching, guiding, and advising in the development of personal and professional identities. More specifically, the mentor plays a key role in initiating the protégée into the world of the mature practitioner in technology (Vetter, 1993).

Feibelman emphasized the importance of selecting a prominent professor to serve as thesis adviser, because such individuals tend to be "part of the 'old-boy network' who can help you survive if times are tough, sometimes even if you don't deserve to" (1993, p. 17). The prominent adviser does not have the need to compete with graduate students: "an adviser who has made his mark views the accomplishments of his students with pride, even joy. They are his research 'children'" (Feibelman, 1993, p. 18). The fact that Feibelman used the masculine pronoun so recently to describe the prominent professor is not to be overlooked. Vetter (1992) pointed out that "women, and particularly those in the physical science, mathematics and engineering fields, were far less likely than the men to be tenured, and their academic rank is as much as a full step less" (p. 17). The same problem plagues women faculty members in technology and female graduate students seeking prominent women as their advisers.

Feibelman (1993) offered suggestions to mentors for the advising process, saying that advisers should make themselves available to students, give real guidance, teach "survival skills," be comfortable talking to beginners, and be part of a network that includes graduate students. For traditional graduate school mentors, providing guidance in these areas to women has required new approaches, if not a new paradigm of mentoring. The different ways men and women communicate (Tannen, 1990) and the different ways they are viewed in the graduate setting in technology have complicated the advising process for both mentor and protégée.

MENTORING WOMEN IN THE TECHNOLOGY PROFESSION

Women who complete doctoral study face overwhelming obstacles in obtaining appropriate professional opportunities in academic settings. Technology, like engineering, science, and mathematics, is a male-dominated field. Although some progress has been made in increasing the numbers of women in the other fields, serious attention has only recently been given to dealing with the problem in technology. Vetter (1992) cited a Catalyst (1992) survey of women engineers that "lists exclusion from the Old Boy Network as an important impediment to advancement. Networking allows engineers and managers to learn about work-related matters, as well as the all-important rules of how to play the corporate 'game' to succeed" (p. 12).

Both networking and mentoring were among the strategies associated with effective programs for women in science and engineering in the United States Department of Energy (DOE, 1991). Markert (1996) emphasized the importance of a strong mentor for women in technology. She pointed out that "mentors are critical (male or female) to insure that a younger person's accomplishments are publicized and showcased for management's attention. It is difficult for women to demonstrate their talents if they do not have the assistance of a mentor—especially if they don't fit the typical STEM (science, technology, engineering, and mathematics) mold" (pp. 25-26).

The study of high achievers may not be the most fertile way to develop a strategy for attracting more women to technology and science. From what is known about the appropriateness of the mentor-protégée relationship, it may be that moderately successful individuals communicate a more realistic expectation for attainment. In reporting a study of women in engineering, the Society of Women Engineers noted that

> Positive role models, often male relatives, have played an important part in attracting women to the engineering field. However, as female engineers move into middle and upper levels within their companies, they find very few women who have reached these levels and who can serve as their role models or mentors. (U.S. Congress, 1994, p. 73)

IMPORTANCE OF MENTORING WOMEN IN THE WORKPLACE

In 1977, when Harragan wrote *Games Mother Never Taught You*, she included only a short description of the mentoring process in the corporate setting. She assumed that women's mentors in the business world would be men. She wrote that "confident, self-assured, ambitious women can establish honest human relationships with equally confident, self-assured, ambitious men . . . If you can find one of them, you are very lucky. Your chance of winning the game of corporate politics has just multiplied one-hundredfold" (pp. 380-381).

Vetter (1993) noted that "mentors help in the long process of career development; they initiate mentees into the occupational world by introducing them to its formal and informal parts, its values, customs, resources, and players" (p. 272). She wrote that "mentors teach both how to get things done and what not to do" (p. 272).

Becoming a mentor is a major contribution. Kathleen O'Brien, of the Philadelphia law firm of Montgomery, McCracken, Walker & Rhoads, was quoted in *Working Woman* magazine: "Now we have a history of being mentors to one another and helping young associates. What we give them is a safe information channel, a reality check of sorts, on how they're being perceived, who they should be trying to work with and what kinds of cases they should be working on" (Hadley & Sheldon, 1995, p. 202).

Heim and Golant (1992), in reflecting on perceptions of the need for mentors, said that "women often perceive mentors as unnecessary at best and paternal at worst. They may think, 'I do good work. I'll be rewarded on my merits. Why do I need a mentor? What are they good for?'" (p. 65). The consequences of this point of view may be much more substantial than would be evident on the surface. By overlooking the opportunity to take advantage of the experience of a mentor, women "not only . . . miss the importance of interpersonal work for career advancement, but they remain ignorant of the rules of hardball being played around them. Indeed, they may remain blind to the fact that a game exists at all" (Heim & Golant, 1992, p. 65).

Heim and Golant (1992) pointed out that "having a mentor may be more critical to a female's success than it is to her male colleagues'" (p. 65). They cited a study *(Breaking the glass ceiling)* of individuals in the senior ranks of organizations that showed only 38% of successful men had mentors, but *all* women executives identified their mentors. They commented that "a mentor doesn't appear to be optional for women, probably because women are less familiar with how hardball is played" (p. 65). Heim and Golant (1992) emphasized the reflective role of the mentor, saying that "mentors hold up a mirror to help us see ourselves by giving feedback" (p. 65).

It may be necessary to draft a mentor. The shortage of female role models at J. C. Penney was overcome by a continuing series of speeches by successful women (Hadley & Sheldon, 1995). It is important for the work group to "grow its own role models."

The Philadelphia Women's Network sponsored a program that paired mentors and protégées based on the protégée's goals for career improvement. The Network required mentors to have nine years of college education and experience combined. Protégées are typically women who have not yet met that membership requirement (Hadley & Sheldon, 1995).

MYTHS ABOUT MENTORING

In recent years, mentoring has acquired a mystique that often reaches far beyond reality. Although the present state of knowledge about mentoring is far from scientific objectivity, current evidence disputes many commonly held beliefs about mentoring.

Sandler (1993) subjected conventional wisdom about mentoring to a penetrating analytical critique. She cast doubt upon many commonly held beliefs when she described 10 myths about mentoring:

1. Having a mentor is the best way to succeed.
2. Mentors should be older than protégées.
3. A close, intense relationship is the best primary way to learn about one's profession and to move up the ladder.
4. Mentoring relationships must be long-lasting to be truly useful.
5. A person can have only one mentor at a time.
6. Mentoring is a one-way relationship, benefiting only the protégée.

7. Protégés must be invited.
8. When men mentor women, a sexual encounter is inevitable.
9. Men are better mentors for women.
10. The mentor always knows best. (pp. 274-278)

Conventional wisdom about mentoring gives considerable credence to these beliefs, but Sandler has convincing arguments to discredit these generalizations because they are based upon limited perspectives of the mentoring relationship.

MENTORING WOMEN IN TECHNOLOGY

Mentoring evolved in the tradition of the Good Old Boy Networks, where mentors nurtured people similar to themselves. As women entered fields influenced by the traditional Good Old Boy Networks, they found themselves unable to take advantage of mentor relationships. Older, higher ranked men have been reluctant to support younger, less experienced women.

The selection of mentors is a process that should be approached rationally and analytically. Heim and Golant (1992) wrote:

When hunting for mentors, keep in mind the following:

1. It's better to have several.
2. A man and a woman will have different strengths to offer.
3. Choose someone highly respected, someone whom you admire.
4. Your mentor's access to the dominant coalition can be critical.
5. Ideally, you should feel comfortable with your mentor.
6. His or her value system should be similar to yours.
7. He or she should be willing to give you some time.
8. You should be able to get insights from your mentor that you couldn't derive on your own. (p. 67)

Mentors are less likely to be next-in-line supervisors today than in the past. Mentors may come from settings outside the protégée's work environment, from professional groups or former vendors or clients. In today's climate of job mobility, individuals may have a number of mentors simultaneously. RoAne, in *The Secrets of Savvy Networking*, encouraged the use of mentors of the moment (MOMs) rather than relying upon "mentor monogamy" throughout a career (Hadley & Sheldon, 1995).

GUIDELINES FOR THE MENTORING PROCESS

Several contemporary authors provide excellent guidelines for mentoring. One of the best guides is by Sandler (1993), who developed this set of suggestions specifically for women serving as mentors to women:

The (Woman) Mentor's 10 Commandments

1. Almost anyone can be a mentor to someone else. Many women underestimate how much knowledge they have about their system, academic or other, the number of their contacts, and the avenues they can open. One doesn't have to be boss to be a mentor. Teaching assistants can mentor other graduate students, graduate students can mentor undergraduates, undergraduate majors can help those just beginning.

2. Recognize and evaluate what you can offer a protégée, keeping in mind that you cannot and should not fulfill every mentoring function.

3. Clarify expectations about how much guidance you will offer concerning personal as well as professional issues.

4. Give criticism as well as praise when warranted. Always present criticism with specific suggestions for improvement. Do it in a private and nonthreatening context—over lunch or coffee are good times.

5. Where appropriate, "talk up" your protégée's accomplishments to others. Good forums are in the department, organi-

zation, laboratory, or institution, at conferences and at other meetings.

6. Include protégées in informal activities whenever possible. Invite them to lunch, to discussions following meetings or lectures, and/or to dinners at academic conferences.

7. Help protégées learn what kinds of available institutional support junior persons should seek in order to further their own career development. Tell them about funds to attend a workshop, for example, or released time for special projects.

8. Tell your protégée if she asks for too much or too little of your time.

9. Remember that you can't mentor everyone or be all that anyone needs. Work within your institution to develop formal and informal mentoring programs as well as encouraging social networks. Work also to develop ways for information and facts to be shared formally through printed materials and meetings, such as having senior faculty do a panel for junior faculty on how to get tenure.

10. Help not only those who are like you but also those who are not. I have always believed that it is far easier for women than for men to cross boundaries such as race, color, ethnicity, class, and religion. And we can do that when we mentor each other. (pp. 278-279)

Another excellent set of suggestions has been developed by Hadley and Sheldon (1995), who offered four principles to those who would serve as mentors to women in the corporate sector. They wrote

When acting as a mentor:

1. In order to be a good mentor, you must be willing to make a commitment. Mentoring doesn't mean tossing off advice or bits of wisdom when it suits your schedule. Be prepared to invest time and energy into the individual.

2. Listen! Because you're the mentor, you might fall into the trap of doing all the talking. Listening is always important when networking, but particularly in a mentoring situation. You're there to *guide* and *coach*, not lecture and pontificate.

3. Give constructive feedback. Offer specifics whenever possible. Instead of 'You need to work on your negotiating skills,' point to identifiable behaviors: 'When you crossed your arms and pushed your chair away from the table, I saw the negotiations come to a halt. Keep your body language open.'

4. Recognize when it's time to end the mentoring relationship. This could happen for a number of reasons: Your protégée has reached an experience level that surpasses yours, one of you can't commit the time or effort to maintain the relationship, or your mentee's goals have changed. Chances are, you'll know when it's time to say good-bye. There may be a distance or an awkwardness between you, or perhaps you don't make as much of an effort to get together. When it happens, look at it as a positive sign. You've fulfilled a need, and your protégé is now comfortable in pursuing his or her career goals without you. Offer your congratulations and suggest that you continue on as colleagues. (pp. 55-56)

To get the most benefit from mentor-protégée interactions, the protégée also needs to give attention to the most productive approaches. Hadley and Sheldon also developed a set of useful guidelines for protégées to use to maximize the value of their relationships with their mentors. They wrote

When working with a mentor:

1. Don't take your mentor for granted. Be sure to acknowledge your mentor's contributions with thank-you notes, lunches and other forms of appreciation.

2. It may seem perfectly natural to let mentor relationships develop into one-sided affairs. After all, the mentor has all the experience, so it's easy to let him or her do all the giving. But even

if you are not your mentor's equal in experience, skill, or connections, you still have plenty to offer. You undoubtedly will be able to share connections, information, and experiences as well.
3. Respect your mentor's time. Be prepared with specific topics, questions, or problems when you meet.
4. Be willing to accept your mentor's input, even criticism. And be prepared to act on your mentor's advice. (1995, p. 55)

If one broadens the mentoring perspective somewhat to include coaching strategies, the structure of the interaction takes a slightly different perspective. The National Society of Professional Engineers (1992) offered a set of suggestions for effective coaching and mentoring in the corporate setting:

> Coaches/mentors may be inside or outside an organization. An employee may choose a coach/mentor from outside the organization for professional guidance, but select one from inside the organization if employer-related guidance is sought. It is appropriate for men to coach/mentor women, particularly if the guidance focuses on business areas in which there are few women employed . . . Coaching/mentoring is an agreement between two people, not a mandate for a protégée to follow advice or suggestions. Coaching/mentoring offers new insights into general problems, not quick answers to specific problems. Coaching/mentoring is a voluntary undertaking, not an assignment. Coaching/mentoring is not an opportunity for a select few or employees requiring remedial work, but for everyone. (p. 38)

Whatever approach is taken in the mentoring process, it is important to monitor the actual interaction frequently to be sure that it continues to be beneficial to both persons. Some protégées may seek more (or less) mentoring than might be most valuable; some mentors may become overwhelmed by the need to give time and attention to a colleague, or it may be time for the protégée to seek input from other mentors. This monitoring of the mentoring is best done as a part of the ongoing dialogue between the players.

BENEFITS (AND RISKS) TO THE INDIVIDUALS INVOLVED

It seems obvious from the discussion so far that mentoring offers many advantages to both protégée and mentor. Few activities in one's professional life can be so rewarding. However, some areas and developments may not be as positive for either person.

Men face unique challenges when serving as mentors to women. In the reversal of the traditional formula for women's success, "think like a man," men need to hone their awareness of women's perceptions to comprehend the ways they approach a situation. Unless male mentors can "think like a woman" well enough to frame the problem situation accurately from the feminine perspective, it is unlikely that their suggestions will be meaningful or useful to their protégées.

Mentoring, which bestows a plethora of advantages upon mentor and protégée alike, also involves negative outcomes that may surprise the unwary practitioner. Sandler (1993) has cataloged several risks of mentoring and contrasted mentoring with less formal social networking:

> Mentoring relationships are generally exclusive and one-on-one. In contrast, social networks include large numbers of people—all of one's eggs are not in one basket. Thus, one can gain a variety of kinds of information from a social network rather than having it all funneled through one person, the mentor. The more intense the mentoring relationship, the harder to make a necessary break, and the more potential that such a rupture will be disruptive and potentially damaging. The casualness of networking relationships makes beginnings and endings easier.
>
> The mentor-protégée relationship is not casual; it takes time, energy, emotional commitment, and involvement. The

> investment in a mentoring relationship may mean isolation from other equally productive and valuable interactions.
>
> The mentoring relationship typically involves a hierarchy. Mentoring relationships do not take place between equals. Social networks may involve unequals, but they may also include reciprocal exchanges of information. The social network involves people at all levels of the organization—some higher, some lower, some equal—as well as individuals outside it. Anyone who may help advance one's career may participate. John P. Kotter (1982) found that the people judged most effective on the job had the largest social networks.
>
> Mentors often set the agenda, and it is *their* agenda rather than that of the protégée. Mentors decide if they want the role and set the terms of the relationship. In contrast, social networks allow input into what kind of help one needs and how to best find it.
>
> Finally, relying primarily on a mentor for emotional support, as well as for information, evaluation, coaching, and introductions, implies that the mentor has to be superior on all fronts—a hard role for both participants. Support can come from many relationships in one's social network. (p. 274)

The unseen risk, then, is that the person chosen as mentor, who probably has a strong social network, will offer advice based upon information from that network but overlook the importance of helping the protégée build her own strong social network to enhance her perceived job effectiveness.

Clearly, both mentor and protégée need to give careful, thoughtful attention to the mentoring processes if each is to achieve maximum benefit from the relationship. Giving advice may be a natural activity; giving sound, accurate guidance in complicated circumstances over a long time requires considerable wisdom. The protégée must be both discerning and persistent to capitalize upon the suggestions of a mentor and make optimum progress in her career.

SUGGESTIONS FOR THE FUTURE

What stage of progress has been attained in fostering women's status and achievements in technology? Rosser (1995) noted that "the physical sciences, mathematics, and engineering persist as the professional areas where women have not yet broken the gender barrier" (p. 1). Rosser described a six-phase model for curricular and pedagogical transformations in science: absence of women is not noted, recognition that most scientists are male and that science may reflect a masculine perspective, identification of barriers that prevent women from entering science, search for women scientists and their unique contributions, science done by feminists and women, and science redefined and reconstructed to include us all. If Rosser's model is generalizable to the role of women in technology, it could provide a helpful guide to the identification of the present situation and help direct future efforts intended to improve the situation.

Perhaps the major key to success in the mentor-protégée relationship in technology was identified by Fort (1993), who wrote

> Women have to get into more positions of power. But I notice when women do, many of them keep other women out instead of being role models. If women started to serve as mentors to other women, I think we'd see policies start to change. We need more women mentoring women. (p. 47)

Vetter (1993) cited a specific example from science that seems applicable in technology. She said that

> one reason that women find it so difficult to breach the walls of the Academy (National Academy of Sciences, the active membership of which is only 4% female) is that they lack mentors among the members. Individuals must be nominated before being considered for membership. Often, the nominator is their mentor, who then extols their accomplishments to fellow members. Although the majority of successful women in science have had mentors—largely male ones—there are still too few who think first of their

female students or assistants when asked for recommendations. (p. 268)

Mentoring efforts must be strengthened to enhance opportunities for women in technology. Both men and women have responsibilities for providing good mentoring. Successful women in technology have an obligation to serve as role models, provide exemplary mentoring, engage in a wide range of professional networks, and encourage their protégées to new pinnacles of success.

What contemporary advice seems most appropriate for women in technology? "The bottom line for today is: Find your own sage . . . A real mentor works with you, grooms you, fashions you, prods you, pulls and pushes over a period of time. Changes you and clicks you into place" (Popcorn & Marigold, 1996, p. 391).

REFERENCES

Catalyst. (1992). *Women in engineering: An untapped resource.* New York: Author.

Committee on Women in Science and Engineering (CWSE), National Research Council. (1991). *Women in science and engineering: Increasing their numbers in the 1990s.* Washington, DC: National Academy Press.

Feibelman, P. J. (1993). *A Ph.D. is not enough: A guide to survival in science.* Reading, MA: Addison-Wesley.

Fort, D. C. (Ed.). (1993). *A hand up: Women mentoring women in science.* Washington, DC: Association for Women in Science.

Hadley, J., & Sheldon, B. (1995). *The smart woman's guide to networking.* Franklin Lakes, NJ: Career Press.

Harragan, B. L. (1977). *Games mother never taught you: Corporate gamesmanship for women.* New York: Warner Books.

Heim, P., & Golant, S. K. (1992). *Hardball for women: Winning at the game of business.* Los Angeles: Plume.

Markert, L. R. (1996). *Achieving success and influence in science and technology regardless of gender.* Unpublished paper.

National Society of Professional Engineers. (1992). *The glass ceiling and women in engineering.* Alexandria, VA: Author.

Popcorn, F., & Marigold, L. (1996). *Clicking: 16 trends to future fit your life, your work, and your business.* New York: HarperCollins.

Rosser, S. V. (1995). Reaching the majority: Retaining women in the pipeline. In S. V. Rosser (Ed.), *Teaching the majority: Breaking the gender barrier in science, mathematics, and engineering* (pp. 1-21). New York: Teachers College Press.

Sandler, B. R. (1993). Mentoring: Myths and realities, dangers and responsibilities. In D. C. Fort (Ed.), *A hand up: Women mentoring women in science* (pp. 271-279). Washington, DC: Association for Women in Science.

Tannen, D. (1990). *You just don't understand.* New York: Morrow.

U.S. Congress, House Committee on Science, Space, and Technology. (1994). *Careers for women in science and technology: Hearing before the Subcommittee on Energy* (May 12, 1994). Washington, DC: U.S. Government Printing Office.

U.S. Department of Energy (DOE), Office of University and Science Education Programs. (1991). *Review of laboratory programs for women.* Summary of a meeting hosted by Argonne National Laboratory and the University of Chicago, November 16, 1990.

Vetter, B. M. (1992). *What is holding up the glass ceiling? Barriers to women in the science and engineering workforce.* Washington, DC: Commission on Professionals in Science and Technology. (Occasional Paper 92-3.)

Vetter, B. M. (1993). Status of women in technology: Mentors needed. In D. C. Fort (Ed.), *A hand up: Women mentoring women in science* (pp. 267-270). Washington, DC: Association for Women in Science.

Effective Leadership for All

Elizabeth Smith
Pensacola Junior College

Leadership is a topic of extensive research, training, and discussion among educators today. There are as many different leadership styles as there are leaders, gender and ethnicity not withstanding. In this chapter I discuss four leadership styles and three leadership tenets from the perspectives of Sally Helgesen (1990) and Wess Roberts (1985). In the book, *The Female Advantage*, Sally Helgesen (1990) chronicled four women CEOs and their leadership styles. In *Leadership Styles of Attila the Hun*, Wess Roberts (1985) used Attila, a historical barbarian, to relate current leadership theories. In her book, Helgesen (1990) used the spider web as the model for effective leadership, with the leader in the center of the organization and the organization encircling the leader. Roberts (1985) referred to the traditional top-down leader, with Attila at the top; however, the leadership traits are more similar than not. Also included in the chapter are leadership scenarios, both real and imagined. Finally, examples of strategies to use in leadership are given.

LEADERSHIP STYLES

Each person develops his or her own leadership style, based on, for example, personality, childhood, education, and role models. In this chapter, leadership styles are not presented as good or bad, but different. Four leadership styles are charismatic, dictatorial, transactional, and transformational.

Charismatic leaders lead primarily by the strength of their personality. Motivation and empowerment appear somehow magically infused into the people and the organization by charismatic leaders. Communication is open, and information and knowledge are shared. Followers of charismatic leaders own their decisions and are given credit for having the intelligence to make sound decisions for the good of the organization. Diana, Princess of Wales and Franklin Delano Roosevelt are examples of charismatic leaders.

Diametrically opposed to the charismatic leader is the dictatorial leader. The dictatorial leader leads from the top down and from his or her knowledge, not shared knowledge. The followers of the dictatorial leader need no particular intelligence or decision-making capabilities, merely the ability to follow orders and take commands. Although this type of leadership may be viewed as negative, during the heat of battle it is most appropriate. General George Patton and Catherine the Great are examples of dictatorial leaders.

The transactional leader is, as the word implies, a managerial leader. This leader dots every I, crosses every T, and keeps things on track. Reports and statistics abound for this leader. The transactional leader manages the bureaucracy, not the people. With some traits of the dictatorial leader, the transactional leader is secure only when in control of the facts and figures, which are promulgated as an effort to appear in control. Elizabeth Dole and Madeline Albright serve as models of this leadership style.

At the other end of this spectrum is the transformational leader. That person has the vision and foresight to transform people, institutions, or nations. Not unlike the charismatic leader, the transformational leader draws heavily from personality to affect and effect change. Mother Teresa and Lee Iacocca are examples of transformational leaders.

An effective leader has a unique and preferred leadership style; however, he or she also has the ability to use other leadership styles, depending on the time and situation. Hitler may best be recognized for his dictatorial leadership, but he was also extremely charismatic. He could excite crowds of people to mass hysteria. He had the strength of personality to make his followers commit unthinkable atrocities to their fellow humans. Although Elizabeth Dole may be identified as a transactional leader, she has been transformational by using her celebrity status to generate awareness and support for the American Red Cross.

Effectiveness of leadership styles can be seen in all walks of life and professions. A purchasing agent for a nonprofit agency can serve as an example of the effectiveness of a charismatic leader versus a dictatorial leader. Purchasing agents, by the very nature of their jobs, must be transactional leaders. They must attend to detail and ensure that all accounting is accurate and timely. A dictatorial purchasing agent for a large nonprofit organization was very good at her job, but unbearable to work with. Her idea of enforcing the rules was to prohibit her staff from taking bid quotes. She wanted to control every detail and, in doing so, slowed the process to an unacceptable pace. After years of receiving complaints about the authoritarian manner in which this agent worked, she was fired, and a charismatic leader was hired. The charismatic purchasing agent lived and worked by the same rules but empowered the staff to be decision makers. She also had effective communication skills and worked with departments to acquire the needed products. The charismatic purchasing agent enforced the rules as effectively as did the dictatorial purchasing agent but with an open and affirming communication style.

LEADERSHIP TENETS

To be an effective leader, one must follow the tenets or doctrine of leadership regardless of preferred leadership style. Decision making, communication, risk taking, and humor are all keys to successful leadership.

Decision Making

Decision making is one tenet of effective leaders. Helgesen (1990) reported that by operating from the center of the web, the leader has direct access to information that allows wider input into the decision, testing of decisions, and more data to be gathered before the decision is made, but still allows the leader to make a responsible and prudent decision. Requesting input is not tantamount to giving away leadership or authority. Roberts (1985) stated that noble resolve to do the right thing is a characteristic of prudent decision making. Both authors agree that allowing subordinates to make decisions appropriate to their level of responsibility, while accepting full responsibility for these decisions, is a characteristic of an effective leader. Roberts (1985) reminded us that perfect decisions are rare, and responsible decisions are hard to improve on. Open and safe communication is a must for effective leadership that reinforces sound decision making.

Communication

In addition to communication coming down from the leader, the leader must be able to receive communication unilaterally. An effective leader invites both good and bad news. Good listening skills are as vital as good written and oral communication skills. Being approachable and open to accepting good and bad news is a struggle for many leaders, especially transactional leaders. Approachability is not a part of the dictatorial leader's style. The leader's inability to accept bad news places followers in an extremely difficult position.

A strong trait of the transactional leader is that of being "bottom line" oriented. Interpersonal and communication skills are incidental and certainly not a requirement for the transactional leader to accomplish goals. Such was the case of Mr. Jones, a high school principal. Mr. Jones could tell you to the minute the total number of students enrolled on his campus. He could report to the penny his budget situation and could regale you with the success rates of his graduates in post-secondary education. When awards came to the school and pass rates on tests were above expectations, he was quick to report these achievements to the superintendent; however, student or faculty problems were a different matter. Mr. Jones was unable to accept any shortcomings in his faculty, staff, or students. It was as if problematic situations were untenable and a direct reflection of his leadership ability. When his subordinates finally brought problem situations to him, he reacted by assessing blame to the staff and communicated by yelling and bellowing. The coping mechanism adopted by his subordinates was to do everything possible to deal with the problem, rather than to share the problem with him. When it was no longer possible to keep the problem away from Mr. Jones, a true crisis had arisen, which usually necessitated legal counsel for the school district.

Consider this scenario with a transformational leader who pays little or no attention to detail but has great charisma and interpersonal skills. The new leader practices open communication and invites problems as well as celebrations to be shared at the earliest possible time. This new attitude results in significantly fewer and less significant student problems. With open communication and the principal's early involvement in a situation, problem-solving capabilities are magnified. Group discussion by involved persons results in several options for dealing with the situation, all of which the faculty and staff can understand, support, and explain to others. Furthermore, the longer the new principal holds this job, the greater the understanding of the faculty and staff in the principal's decision-making process. As the trust levels builds, and the faculty and staff believe that they will not have to suffer for having problems, the more open they become, and small problems are more easily remedied. Open, safe communication is a must for effective leaders.

Humor

Humor is a communication tool that can be both extremely positive and devastatingly negative. The Attilaism is "great chieftains never take themselves too seriously" (Roberts, 1985, p. 102). Humor differs between human beings. Effective leaders must be able to find appropriate humor in the most difficult situations. Laughter really is the best medicine much of the time. It can provide a connection between people.

Chambers of Commerce and service clubs such as Rotary, Kiwanis, Civitan, and the Masonic Lodge understand the power of humor. When I was invited to join one of these clubs, I was both honored and intrigued. I was not the first female to join, but one of the first four. The membership breakdown is approximately 75 men, 12 women, with an average age of 65 years, and at least a third of the membership consisting of retired military officers.

I joined the club during the presidencies of two younger men (in their 40s). After my first year of membership, I debated strongly about continuing my membership. The experience met none of my needs—professional, personal, nutritional (noon meetings with the standard lunch of fried chicken every week), or social. I was welcome to attend, but I was not included. My club is proud of its reputation of being the most raucous and outrageous club in the city. The young presidents had, for the most part, ended telling racist and sexist jokes, but humor maintained its rightful place of honor in the club. I volunteered to give a program, "Humor in the Workplace." I have presented this program to many different groups and thought it would be of interest to the membership. One major point of this presentation is that humor should be appropriate, timely, and tasteful. At the end of the program, which was well received, a male member asked if I could give an example of inappropriate humor—and I did, by telling a sexist joke about men. They all roared with laughter! I was immediately one of them. I had survived the rite of passage by having the audacity to give back to them the humor of which they were so proud! The lesson learned—if you are not man enough to join 'em—don't.

Did I sell out women? Did I betray our unspoken conviction to be the conscience of the world? No, I did not. I simply allowed my outgoing, outrageous, well-developed sense of humor to be part of my role in that club. By dishing out the same medicine that they dished out, I communicated to them that I was not fragile, not a Southern belle who was easily offended, and that I had a sense of humor. Shortly after my humor program, I was asked to serve as a director on the Board, filling the position that another woman had held. During my first year on the Board, the club changed location, the food improved significantly, and the membership started growing with many more women and younger members.

Risk Taking

"Chieftains . . . must have courage. They must be fearless and have the fortitude to carry out assignments given them (and the)—gallantry to accept the risks of leadership" (Roberts, 1985, p. 17). "The greatness of a (leader) . . . is measured by the sacrifices he is willing to make for the good of the nation" (Roberts, 1985, p. 102).

Every decision involves risks. Risk taking is a vital aspect of effective leadership that makes the difference between greatness and mediocrity. Goal setting also involves risk taking. To set goals, I ask the following questions:

1. What do you want?

2. What is it worth (risk)?

3. How are you going to get it?

Achieving goals in life requires that some aspect of one's life suffers to attain the goal. Helgesen (1990) maintained that men historically sacrifice family for career, whereas women historically balance career and family. I have always considered risk taking to be one of my personal strengths. When I decided to go to graduate school to pursue my Ph.D., I resigned my job, withdrew all of my retirement, borrowed enough money to become a medical doctor, and went to school at Texas A&M University full-time for two years. The risk of unemployment after completion of my doctoral work was worth it for me to finish my program in two years. For me, that risk proved to be the right decision.

Early in my tenure at my current college, I was responsible for the continuing education function, which included non-credit and business and industry programs. At that time, the Downtown Improvement Board was clamoring for the college to establish a teaching site downtown. At the same time, the senior citizens' art class (Art "Grannies") were being evicted from the campus because of the lack of classroom space. I proposed that the college open a downtown center to provide classes for the business community and a place for this community art program, which had more than 200 people enrolled. My offer to the college administration was that if the center was not self-supporting by the end of three years, it would close. The risk paid off. We broke even at the end of 18 months, and 8 years later moved into a larger facility. Leaders, especially women, must learn to take risks and to make the best out of the outcome of those risks.

SUMMARY

The discussion of leadership styles and tenets of leadership in this chapter is only a brief review of this extensive topic. Effective leaders pull from research, education, and experience to provide good leadership. Women are encouraged to participate in leadership training programs and to take calculated risks in their profession. Leadership requires sacrifice and tenacity. It is not for the faint at heart.

REFERENCES

Helgesen, S. (1990). *The female advantage: Women's ways of leadership.* New York: Doubleday.

Roberts, W. (1985). *Leadership secrets of Attila the Hun.* New York: Warner Books.

Environmental and Climate Challenges in Technology Education

Jane A. Liedtke

Illinois State University

IN THE WORLD THAT IS COMING, IF YOU CAN'T NAVIGATE DIFFERENCES, YOU'VE HAD IT.

(Robert Hughes, 1992)

"Progress through the pipeline hasn't been good for women and minorities in the last 12 years. Presidents Reagan and Bush put affirmative action on the back burner, signaling that it wasn't important" contended Galagan (1993b, p. 30). Some government estimates indicate that minorities will make up nearly half of the U.S. population by 2050. Dovidio (1993) reported that in the workplace "racial issues are already receiving serious attention, with companies striving for greater awareness and deeper understanding" (p. 51).

Today, women, people of color, and immigrants hold more than half the jobs in the United States. But the elimination of thousands of middle-management jobs has wiped out much of the proving ground for minorities and women. R. Roosevelt Thomas, of the American Institute for Managing Diversity, argued in *Differences Do Make a Difference* that "corporations do not benefit from the full productive potential of some of their most able employees" if women and minorities are "limited by the continuing 'norm' of white able-bodied males as the ideal" (Galagan, 1993b, p. 31).

"For all the good that they do," wrote Galagan (1993b, p. 30), "diversity programs have yet to make a dent in what most experts consider to be the number-one impediment to advancement for women and minorities: lingering and deep-seated prejudice." Myths about minority groups are part of what holds them back. A popular myth about women, for example, is that the environment will get better for them in the workplace in due time. A report called

"Empowering Women in Business" by the Feminist Majority Foundation in Washington, DC, refuted this and other myths about women in business. According to Galagan (1993b), the report included evidence that equality at the top is *not* just a matter of time. According to the report, "it will take 475 years for women to reach equality with men in the executive suite" (p. 30).

As of 1993, women made up only 2.5% of top executive officers in Fortune 500 companies. More than half of the board chairmen of the Fortune 500 companies are the sons of former chairmen. According to the Associated Press (1996, p. C1), "about 100 of the Fortune 500 companies have no women corporate officers at all." In addition, a report by Catalyst (Associated Press, 1996) indicated that in 1995 only 1,303 of the top corporate officers (n = 12,885) were women. And, of the 2,500 top wage earners, only 50 were women. According to Catalyst President Sheila Wellington, "there is still a glass ceiling, but equally important, this census documents the existence of glass walls" (Associated Press, 1996, p. C1). "Time alone will not cure this matter of women advancing to the top . . . We need vocal, sustained commitment from the top," said Wellington (p. C1).

"These and other hard-wired attitudes are behind the policies and practices that systematically restrict the opportunities and rewards available to women and people of color" contended Ann Morrison, co-author of *Breaking the Glass Ceiling* and *The New Leaders: Guidelines on Leadership Diversity in America* (Galagan, 1993b, p. 32). Ann Van Eron, principal of Potentials, an organizational development firm, noted that "diversity is likely to breed tension, conflict, misunderstanding and frustration unless an organization develops a culture that supports, honors, and values differences" (1995, p. 53).

An examination of the workplace and the conditions for women and minorities in business and industry will aid us in technology education as we examine the environment for women and minorities in our profession. Granted, the educational system is somewhat unique and often removed from the private sector, but societal problems rarely exclude themselves from either venue.

THE TECHNOLOGY LEARNING ENVIRONMENT

Imagine a place where very little in the environment looks familiar. The language is different, and everyone or almost everyone around you looks different. That's the scenario for women and minorities when they enter many technology education programs. The room is probably filled with devices not seen at home or used previously in school. The terminology is so new and different that few words relate to daily life. There is also the isolation of being the "only child"—the only girl, the only Asian or Asian-American, the only African-American, etc. When a young woman or minority enters a technology education program, what does she or he experience?

To increase the enrollment of women and minorities in technology education at all levels, we must make the environment welcoming, supportive, and conducive to making the transition from the unfamiliar to the familiar. This requires technology educators to focus on their own awareness of the classroom/laboratory environment and to analyze carefully what restricts the entry and retention of women and minorities. Being different is not the "problem" of the young African-American girl who is captivated by lasers and robotics. Thus, she must not be forced into the mold of how White men learn about and gain experience with lasers and robotics. The educational environment must change to accommodate her and provide experiences with technology that relate to her daily life experiences. This accommodation requires curriculum innovations, new pedagogical approaches, and often a real change in how the instructor perceives and interacts with students.

CHALLENGES IN TECHNOLOGY EDUCATION

Public school systems throughout the United States engage thousands of students annually in the study of technology through technology education programs. Many students are college bound. These technologically literate graduates have many career options: business, communications, engineering, graphic design, industrial technology, medicine, physics, etc. Because of a wide range of interests, perceived lucrative salaries, or lack of awareness of technology education as a career, few students elect to major in technology education in college. A classroom teacher may find few students and their respective parents excited about the career of teaching technology. This problem, among other factors, has created a major shortage of technology teachers nationwide. Today technology teacher educators must recruit not only from high school technology education programs but from community colleges and undeclared majors already on campus. Of the available students, a few women and minorities each year are compelled by high technology and the prospect of teaching children. They enter into technology education thinking they will have a successful career. If they are lucky, a professor will mentor them through their college program. If not, they likely will encounter experiences with peers or faculty members that are inconsistent with the way they perceive they should be treated. Without a support system in place, these experiences begin to erode the student's ability to cope with the environment. Although the technology education program faculty may not intend to sustain a negative environment for women and minorities, it may still exist. In fact, it may even be invisible to the faculty yet obvious to the underrepresented groups enrolled.

Initial negative experiences cause many high-potential students from underrepresented groups to remove themselves from a program. The "make or break" practice of expecting women and minorities to assimilate causes some students to find other majors or even drop out of school. The need to excel beyond the "average" to be considered acceptable is also problematic. Essential for the retention of underrepresented groups are support systems and faculty who have a commitment to diversity and non-biased educational experiences. Student leadership activities for all technology education majors are also required. Technology teacher educators must also agree that what needs to be "fixed" is the system, which allows discriminatory practices to survive. Women and minorities who make it through the higher education system in technology education should be able to say that they succeeded because of their professors, not in spite of them.

The graduating teacher who happens to be a woman and/or minority is met with two scenarios when searching for a teaching position: the school system that is excited to attract and hire a new teacher from an underrepresented group, or the school system that won't hire a new teacher from an underrepresented group (for whatever reason—e.g., perception of less ability, dominance of White men in the hiring process). The wise graduate takes a position at a school like that described in the former scenario, which seems most supportive of her or his teaching abilities and interests. What may result, however, is that the upper-level administrators who hired the new teacher are unaware or detached from the environment the teacher will encounter daily. That is, the manner in which the new teacher is initiated in the workplace by her or his new colleagues may not be what was promised during the hiring process.

Career induction can be a wonderful inclusionary experience, or it can be absolutely isolating. The new teacher may be hired into a group of technology teachers who have worked together for years, creating the possibility that being new and "different" increases the difficulty of "fitting in." The new teacher may be the sole technology teacher in the school, thus having the burden of being inducted into teaching by those who may not have an appreciation of technology education or what it is like to be a member of an underrepresented group.

Of course, not all situations are negative, and not all new teachers have a difficult time adjusting to the demands of teaching and their work environment. The point is that the environment can't be forgotten, and its influence on the retention of technology teachers from underrepresented groups cannot be overemphasized. The mere fact that we often don't realize a problem exists or we have created what is termed a "null" environment (an environment lacking in encouragement or support and thus restraining performance) is cause for examination of our practices and support systems.

Technology teachers from underrepresented groups have a keen sense of what has happened to them throughout their careers. What keeps them going is the hope that the environment will get better. Maintaining a positive view over the long haul isn't easy.

After repeated negative experiences it's amazing that women and minorities remain in our profession. They do because of positive influences on their careers, the students who have touched their lives, the mentors and role models who have guided them when times were tough, and perhaps a persistent sense of dedication to the teaching profession.

ORGANIZATIONAL CULTURE

Technology education programs and even their discrete classes have a culture all their own. Departments, school buildings, and institutions have idiosyncratic characteristics. Professional associations and professions at large form their own operational style and environment. All represent what is termed organizational culture. Organizational culture has been defined as "the pattern of shared values and norms that distinguishes an organization from all others" (Higgins, 1994, p. 461). These values and norms provide "direction, meaning, and energy for members of the organization," according to Higgins (p. 462). Every organization and profession has a culture of its own that evolves as the members of the organization and their expectations change.

To increase the participation of women and minorities in technology professions, organizational cultures must be attractive to these individuals and consistent with the factors (values and norms) with which these individuals can best identify. To do this, technology organizations such as educational institutions, professional associations, and business/industry must develop and reinforce organizational cultures that ensure and value diversity.

What kind of culture works best? According to Meares (1986), the specific needs of organizations differ. But in general, cultures should be created and managed such that they

- Create and meet employee expectations
- Communicate desired values and beliefs
- Promote interdependence and mutual trust and respect
- Provide mentorship

- Sponsor advantageous directives and philosophies
- Encourage individuals to share their efforts and ideas freely (p. 38)

According to Deal and Kennedy in *Corporate Culture: The Rites and Rituals of Corporate Life* (Goldstein & Leopold, 1990, p. 35), "employees attain yet another sense of who they are, what they should be doing and how they should behave through identification with their organization's culture. The company benefits from this cultural cohesiveness, which is essential for smooth work flow, productivity and a common sense of affiliation that, in turn, contributes to the organization's values and goals . . . Mike Fenton, manager of affirmative action and human resources planning for AT&T's Bell Laboratories, says that people must be comfortable with each other to work well together."

Social scientists examine and describe organizational culture through four kinds of artifacts: myths and sagas; language systems and metaphors; symbols, ceremonies, and rituals; and identifiable value systems and behavioral norms. To focus on the relationship between underrepresented groups and organizational culture, we must become aware of the artifacts that currently define technology education organizations.

Myths and sagas reveal important historical facts about early pioneers and products, past triumphs and failures, and the visionaries who have transformed the profession (Higgins, 1994). These myths and sagas "identify the organization's shared values and norms and reinforce them" (p. 462).

What myths, sagas, stories, and history should be shared about the inception and growth of technology education? Such stories help to shape the attitudes and behavior of new and veteran members of an organization. If the myths and sagas of technology education include women and minorities, then the profession will be seen by new and potential members as one where all can succeed and gain self-esteem. If they can see that performance is rewarded and that the members of the profession care about them, the motivation for women and minorities to belong to the profession will be high.

Language systems and metaphors used in an organization also indicate shared values. How people refer to each other in professional settings and the language used can create the feeling of an open group or a closed society. Male-oriented language in technology education conveys subtle messages to women that may not be supportive of their inclusion. For example, calling someone by their last name is a masculine way to address them and is an unprofessional approach. Referring to members of a group as "guys" is another common masculine classification.

Symbols, ceremonies, and rituals reveal what is important to us. Symbols, logos, flags, and slogans convey the importance placed on certain ideas or events. Mottoes convey much about organizations and serve verbally as a symbol. According to Hersey (Higgins, 1994, p. 463), a good motto meets the following criteria:

- It conveys and promotes the organization's core philosophy.
- It has an emotional, rather than rational or intellectual appeal.
- It is not a direct exhortation for loyalty, productivity, quality, or any other organizational objective.
- It is mysterious to the public but not to members of the organization.

Value systems and behavioral norms are reflected in the profession's strategy, structure, systems, style, staffing, skills, politics, rules, and procedures (Higgins, 1994). Values and norms are passed on in informal communications and can also be seen in what is rewarded. How organizations are structured and the extent to which individuals are allowed to participate in decision making are a critical component of the value system.

Kilmann (1985) found that norms, or informal standards of behavior, play an important part in establishing an organization's culture. About 90% of organizational norms have negative connotations. Findings by Kilmann suggested that culture, as expressed in norms, could have a negative effect.

IDENTIFIABLE CULTURES

Deal and Kennedy (1982) suggested four cultural categories: tough-guy/macho culture, bet-your-company culture, work hard/play hard culture, and process culture. Dr. Jeffrey A. Sonnefeld, Director of the Center for Leadership and Career Change at Emory University Business School, described four kinds of corporate culture: "the Academy, the Club, the Baseball Team, and the Fortress" (Strugatch, 1990, p. 206). These categories serve to describe the work environment and provide definition for why women and minorities may find some cultures foreign to them.

Tough-guy/macho cultures are characterized by highly competitive situations with high-risk strategic decision making. Leaders in this type of culture are "heroes," slogans are "battle cries," and "ceremonies focus on problem solving" (Higgins, 1994, p. 467).

Bet-your-company culture "results from decisions for which feedback is slow but risks are high" (p. 467). The culture is common in capital-intensive areas where major investments are made in technology and equipment yet the payoffs are not known for some time. The "heroes" in this culture are wise and experienced because they have "survived over the long haul" (p. 467) and know what's involved in believing in the organization. The ceremonies associated with this type of culture are formal meetings designed to reduce uncertainty.

The work hard/play hard culture "emerges in situations characterized by fast feedback and low risk" (Higgins, 1994, p. 467). It is considered to be a fast-paced and fun culture where there is plenty of action for everyone and creative problem solving is encouraged. Conventions, meetings, contests, and parties all reinforce the values of hard work and hard play.

The process culture "evolves from situations in which feedback is slow and risk is low" (Higgins, 1994, p. 468). The term *process* refers to how problems are solved and decisions made. According to Higgins, "the key value is the way in which decisions are made—that is, the process" (p. 468). Organizations with a process-oriented culture are often described as mechanistic. "Heroes" in this culture "devise new processes and perform maintenance roles for the organization" (p. 468), and they keep the organization going by passing on information. Ceremonies reward performance in carrying out the process, like 10-year or 25-year awards.

The four categories postulated by Sonnefeld offer some additional possibilities (Academy, Club, Baseball Team, and Fortress) and insights into the personalities that are attracted or best suited to different organizational cultures. The Academy directs organizational members to specialize and celebrates the value of personal training (or level of educational attainment). In the Academy, hierarchy is valued, and movement is vertical. The Academy encourages specialization and long-term commitment.

The Club values versatility and helps organizational members be team players or "family." In the Club, conformity is valued. Both the Academy and the Club tend to attract individuals who value stability, enjoy a variety of challenges, and find they "shine" in group settings (Strugatch, 1990). In both settings, individuals know how to fit in quickly and are participative.

According to Strugatch (1990, p. 207), "if the Club is an extended family, the Baseball Team is a one-night stand. You have to hit a home-run the very first week of the season or you're history." The Baseball Team values those who produce at all costs, even in a high-pressure environment with short-term results.

The Fortress exists in a permanent atmosphere of crisis, expecting organizational members to thrive on it. Like the Baseball Team, the Fortress is for individualists, independent thinkers, and those with little regard for conventional wisdom. The Fortress stresses instinct over training. The heroes at the Baseball Team and Fortress are risk takers guided by their instinct and savvy.

These models from business and industry also relate to the cultures found in educational institutions and professional associations. To learn more about our organizational culture, we might ask these questions: What is the culture of the institution where I am affiliated and the culture of the professional associations in which I hold membership? What context(s) have been created by design or by default that influence people in these organizations and thus contribute to the dilemma of how to increase the involvement of minorities and women?

What Happens to Underrepresented Groups in These Cultures?

It is generally believed that people usually accept (or gravitate toward) organizational cultures that are like themselves and where they can feel comfortable and contribute. Researchers using questionnaires have recognized that many women have entered technology professions because of experiences they had early in life with their fathers (e.g., building or making things, or engaging in technological activity in a positive setting) or the influence/recommendation of their teachers. Somehow through experience in prior situations the "culture" was internalized as being attractive. If this is true, then we in technology professions must offer a broader range of contexts for which individuals of diversity can "fit" by changing the organizational culture, or we must attract those individuals with the personalities and interests that "fit" the organizational culture.

Recruitment is often seen as the means by which individuals can be brought "into the fold." In actuality, recruitment will only be beneficial if retention issues are addressed. For example, recent reports in the press have reported on women in science and engineering careers. An article from Knight-Ridder News Service reported that "women are leaving careers in science and engineering at almost double the rate of men and face a work environment with unequal pay, sexism and few accommodations for family demands, according to a National Research Council Report" (1994, April 3, p. C2). The report, "Women Scientists and Engineers Employed in Industry: Why So Few?" indicated that the reasons for underrepresentation include "an old boy's network that prevents women from finding out about choice jobs. Paternalistic attitudes keep women from getting career opportunities. And hostile superiors who place unreasonable hurdles on women seeking career advancement" (p. C2).

According to Delatte and Baytos (1993), if an organization intends to respect individuality, the underlying assumption is that people new to the organization must go through an assimilation process. "Through this process those who are different are welcomed into the organization but then expected to blend in—or alter their attitudes and behavior to suit the organization's homogeneous culture or management style" (p. 56). They also reported that many people "are growing increasingly dissatisfied with the assumption that adaptation is completely their responsibility or that there is value in only one style" (p. 56).

In some organizations token women are placed in positions of authority or influence by men who may believe that they are "doing the right thing." This weakens the relationships among professional women (Ely, 1994). In Ely's research on the effects of organizational demographics and social identity on relationships among professional women, she cited Kanter's (1977) analysis of the "queen bee syndrome" as being problematic: "Queen bees are token women in traditionally male-dominated settings whom male colleagues reward for denigrating other women and for actively working to keep other women from joining them" (Ely, 1994, p. 207).

In addition, Ely also reported "that white men's extreme overrepresentation in organizational positions of authority have a negative impact on women and nonwhite subordinates" (p. 207). Similarly, Ridgeway (1988) suggested that the disproportionate representation of men over women in senior organizational positions may highlight for women their limited mobility and reinforce their lower status as women, even in work groups composed entirely of women. "When this occurs, women form lower expectations for the positions women, and they as women, are likely to achieve in the organization" (Ely, 1994, p. 207).

Research by Morrison, reported in an interview by Galagan (1993b), indicated that the turnover rate for high-potential women is much higher than for high-potential men. Morrison reported that a common reaction to high turnover among women is to change the benefits package, because men believe that women leave to start families. "Thanks to research by Vicky Tashjian, we know that only 7 percent of female professionals and managers leave for family reasons. Of the rest, 73 percent leave because they see limited career opportunities for women in their companies" (p. 42).

Monsanto recognized it had an organizational problem when it discovered it had poor retention rates among people outside the mainstream culture (Galagan, 1993c). Monsanto found that minorities and women were leaving almost twice as often as White men. In the case of minority women, the rate was three to four times the rate for White men. Monsanto examined its structure and culture, which resulted in the identification of eight process barriers that prevented people from understanding diversity:

- Denial of issues
- Lack of awareness
- Restrictions on bringing bad news up the line
- A lack of trust about how others will perceive and respond to diversity issues
- The need to be in control in all areas of one's job
- A compulsion to fix "them" rather than "us"
- An issue outside one's reality
- Past, well-intended, diversity actions (p. 49)

Galagan (1993c) reported that a common reaction to difference is to "fix" the person whose behavior is different. Thomas Cummins, Diversity Development Director for Monsanto, said, "male managers send women to assertiveness training, hoping they will come back able to make their points and ask for things 'the way a man does'" (p. 49). Cummins said that, in an organization mostly made up of White men, "white-male reality is like water to fish—natural and invisible. So we can't understand why people are doing all these strange things to survive in an environment that's so comfortable for us" (Galagan, 1993c, p. 49).

According to Marshall Singer, author of *Intercultural Communications: A Perceptual Approach* (Goldstein & Leopold, 1990), "when surrounded by the so-called majority, people who belong to an ethnic or other minority group usually are unable to forget their minority identities. Internal as well as external conflicts may arise" (p. 85). Singer also noted that one part of women and/or minorities "argues for assimilation; the other side may resist, perhaps by expressing even stronger links to the minority identity group. In such a situation, ethnic identity can become more, not less pronounced" (p. 85). Ignoring our differences discounts our uniqueness as individuals.

CREATING A CULTURE CHANGE

According to Porter and Parker (1992, p. 45), "(o)rganizations which do not change will not survive." Increasing attention has been given to changing how work is done within the organization. When the fundamental ways in which work is done are changed, new strategies, structure, workforce, technologies, customers, and financial engineering become institutionalized, not idiosyncratic (Porter & Parker, 1992).

Changing the demographics of an organization is an important first step. But effective change will not occur unless diversity exists at all levels, a sound conceptual plan is in place, and the plan is supported vigorously (Anderson, 1993). Anderson contended that "different people feel differently about their roles in an organization, about the ways in which they can contribute, and about the recognition and rewards they receive" (p. 60). Anderson believed it is helpful to think about employees in terms of four styles: learning, human-relations, motivational, and communication. This effort will be more successful when processes "foster equity, consensus, and empowerment" (p. 60).

Is diversity really one of the key issues? Is our lack of diversity what causes the culture to change so slowly? When we recognize the value in diversity, perhaps our attention can be focused toward productive activity that will lead us to achieving the goal of enhanced participation for women and minorities in technology education.

There are many reasons why organizations are attending to issues of diversity. Rossett and Bickham (1994) reported that compliance, harmony, inclusion, justice, and transformation are all part of the diversity puzzle for organizations. "Compliance" reasons focus on legal aspects of equal opportunity, including racial and sexual discrimination. "Harmony" includes the desire to have people get along with one another and to appreciate each other. "Inclusion" targets underrepresented groups and helps members work with diverse colleagues. "Justice" eradicates the lack of efforts to correct for lack of diversity in the past. "Transformation" means changing the values, processes, and standards of organizational behavior.

According to Rossett and Bickham (1994, p. 41), "(m)any seem to plunge in without giving much thought to their specific goals. At the most basic level, some organizations don't consider whether their purpose is to change individuals or the organization or both." Delatte and Baytos (1993) cautioned that efforts to diversify are "unlikely to be particularly effective if they are conceived and designed via the BOWGSAT method—that is, a Bunch of White Guys Sitting Around a Table" (p. 59). To respond to the concerns and issues of diversity there needs to be diverse input.

In 1990, R. Roosevelt Thomas, Director of the American Institute for Managing Diversity at Morehouse College in Atlanta, suggested 10 steps for managing cultural diversity successfully so that no members of the organization experience an unnatural advantage or disadvantage (Higgins, 1994, p. 476):

1. Clarify your motivation.
2. Clarify your vision.
3. Expand your focus.
4. Audit your corporate culture.
5. Modify your assumptions.
6. Modify your systems.
7. Modify your models.

8. Help people pioneer.
9. Apply the special consideration test.
10. Initiate affirmative action.

Thomas contended that the managerial environment must change and help people to understand that a culturally diverse organization enables everyone to contribute to her or his greatest potential. Managing cultural diversity, said Thomas, "goes beyond integrating minorities and women into the work force. Its goal is to create a heterogeneous culture in which people differ in many ways, including age, education, background, function, and personality" (Higgins, 1994, p. 476).

To create the desired culture, information about the existing culture is needed, and the organizational culture's real values must be changed to meet the requirements of the new culture (Thomas, 1990). In the change process, members of the organization must be helped to "overcome obstacles and recover from failures" (Higgins, 1994, p. 476).

Bailey Jackson of the University of Massachusetts developed four basic principles that can be used to identify progress toward a multicultural organization (Jackson, LaFasto, Schultz, & Kelly, 1992). A multicultural organization

1. Reflects the contributions and interests of diverse cultural and social groups in its mission, operations, and product or service

2. Acts on a commitment to eradicate social oppression in all forms within the organization

3. Includes the members of diverse cultural and social groups as full participants, especially in decisions that shape the organization

4. Follows through on broader external social responsibilities, including support of other institutional efforts to eliminate all forms of social oppression. (p. 22)

Jackson and Hardiman (Jackson et al., 1992, pp. 22-24) identified stages that an organization may go through. Figure 1 illustrates these stages that an organization may move through as it becomes a multicultural organization.

Level One

Stage One: The Exclusionary Organization

The Exclusionary Organization is devoted to maintaining dominance of one group over other groups based on race, gender, culture, or other social identity characteristics. Familiar manifestations of such organizations are exclusionary membership policies and hiring practices.

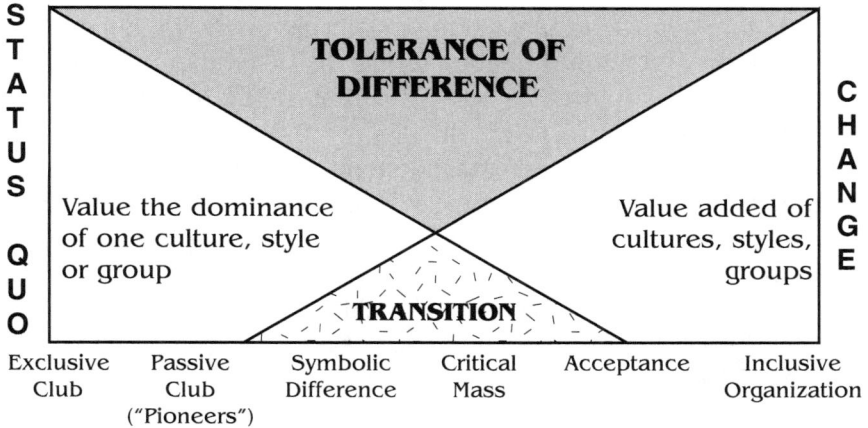

The Path from a Monocultural Club to a Culturally Diverse Organization*

Judith H. Katz and Frederick A. Miller

*This model was originally developed by Bailey Jackson, Rita Hardiman and Mark Chesler (1981) "Racial Awareness Development in Organizations" and adapted in 1986 by Judith H. Katz and Frederick A. Miller, The Kaleel Jamison Consulting Group, Inc.

Stage Two: The Club

The Club describes the organization that stops short of explicitly advocating anything like White male supremacy, but does seek to establish and maintain the privilege of those who have traditionally held social power. This is done by developing and maintaining missions, policies, norms, and procedures seen as "correct" from their perspective. The Club allows a limited number of members from oppressed groups such as women and racial minorities, provided that they have the "right" perspective.

Level Two

Stage Three: The Compliance Organization

The Compliance Organization is committed to removing some of the discrimination inherent in the "club" by providing access to women and minorities; however, it seeks to accomplish this objective without disturbing the structure, mission, and culture of the organization. The organization is careful not to create "too many waves" or to offend or challenge its employees' or customers' racist, sexist, or anti-Semitic attitudes or behaviors.

The compliance organization usually attempts to change its organizational racial and gender profile by actively recruiting and hiring more racial minorities and women at the bottom of the organization. On occasion, they will hire or promote "token" racial minorities or women into management positions, usually staff positions. When the exception is made to place a woman, racial minority, or member of any other oppressed social group in a line position it is important that this person be a "team player" and that s/he be a "qualified" applicant. A "qualified team player" does not openly challenge the organization's mission and practices and is usually 150 percent competent to do the job.

Stage Four: The Affirmative Action Organization

The Affirmative Action Organization is also committed to eliminating the discriminatory practices and inherent "rigged" quality of The Club by actively recruiting and promoting women, racial minorities, and members of other social groups typically denied access to our organizations. Moreover, the affirmative action organization takes an active role in supporting the growth and development of these new employees and in initiating programs that increase the chances of success and mobility. All employees are encouraged to think and behave in a non-oppressive manner, and the organization may conduct racist and sexism awareness programs toward this end.

This organization's view of diversity also includes the disabled, Latinos, Asians/Asian American-Pacific Islanders, Native Americans, the elderly, and other socially oppressed groups.

Although the affirmative action organization is committed to increasing access for members of diverse groups and increasing the chances that they will succeed by removing those hostile attitudes and behaviors, all members of this organization are still required to conform to the norms and practices derived from the dominant group's world view.

Level Three

Stage Five: The Redefining Organization

The Redefining Organization is a system in transition. This organization is not satisfied with being just "anti-racist" or "anti-sexist." It is committed to examining all of its activities for their impact on all members' ability to participate in and contribute to the growth and success of the organization.

The redefining organization begins to question the limitations of the cultural perspective as it is manifest in its mission, structure, management technology, psychological dynamics, and product or service. It seeks to explore the significance and potential benefits of a diverse multicultural workforce. This organization actively engages in visioning, planning, and problem-solving activities directed toward the realization of a multicultural organization.

The redefining organization is committed to developing and implementing policies and practices that distribute power among all of the diverse groups in the organization. The redefining organization searches for alternative modes of organizing that guarantee the inclusion, participation, and empowerment of all its members.

Stage Six: The Multicultural Organization

The multicultural organization reflects the contributions and interests of diverse cultural and social groups in its mission, operations, and product or service; it acts on a commitment to eradicate social oppression in all forms within the organization; the multicultural organization includes the members of diverse cultural and social groups as full participants, especially in decisions that shape the organization; and it follows through on broader external social responsibilities, including support of efforts to eliminate all forms of social oppression and to educate others in multicultural perspectives.

Carr (1994) indicated that "individuals vary widely in their openness to and enthusiasm for change" and that "the person most comfortable with any particular change is the one proposing it" (p. 55). People resist being changed—especially when the change appears to have a payoff primarily for someone else. Carr contended that in order to change we must "understand the factors that matter in change and what impact they have on the people we expect to change" (p. 56). He proposed and described seven key factors (or questions) that play a role in the change process:

1. Is this change a burden or a challenge?

    "A change with a clear payoff for those who must do the changing will feel like a challenge. If it lacks such a payoff it will feel like a burden" (p. 56).

2. Is the change clear, worthwhile, and real?

    "If an organization presents a proposed change so that its benefits appear unclear, trivial or highly unlikely to materialize, the change almost certainly will be seen as a burden to be avoided. On the other hand, when the change promises clear, worthwhile and believable benefits, it will look desirable" (p. 56).

3. Will the benefits of the change begin to appear quickly?

   "The longer a change takes, the hazier its payoff will appear and the more it will seem a burden" (p. 56).

4. Is the change related to one function or a few closely related functions?

   "Nothing is more dear to the units of a traditional organization than preserving their functional integrity. The more functions that must cooperate to produce a change, the greater the probability that at least one function will see itself as a loser in the change and work to sabotage it" (p. 57).

5. What will be the impact on existing power and status relationships?

   "Many organizational players work assiduously to accumulate power and status. Even those with other goals normally appreciate having power and status. And those who have it unfailingly work to maintain it. If a change directly attacks the power and status of any function or group, those who profit from the established situation will certainly oppose it, overtly or covertly. The more that a proposed change conforms to the existing power and status structure, the less likely it is to be opposed by entrenched powers" (pp. 57-58).

6. Will the change fit the existing organizational culture?

   "Transformational changes fail more often than they succeed. Even when they're successful, the cost to the organization is always high and the payoff may be considerably less than expected. Furthermore, major changes almost always succeed only because the organization is facing a major crisis. Is the survival of your organization at stake? If not, then the better the change you propose fits the values of the existing culture, the better the chances of success. Even sweeping changes can be based on core values of the organization" (p. 58).

7. Is the change certain to happen?

"People are much more likely to get involved or to go along with something if they believe it is really going to happen. The point is simple: If you want something to change, line up enough organizational horsepower to ensure that it will—before you start the change" (p. 58).

Because culture is often hard to define or articulate, it is usually difficult to develop practical approaches for changing an organization's culture. According to Craig (1993), even though we cannot change culture directly, we can use the elements of organizational design as levers. "By pushing these levers the right way, a company can create new attitudes and behaviors" (p. 16). These levers are

- Organization structure—the formal relationship between workers
- Work processes—activities linked to accomplish a task or to produce a product
- Management and information processes—the vision, goals, and tasks of the organization and measurements of what employees are doing to meet these goals and tasks, including pay, incentives, and other rewards; planning; training; and formal and informal methods of communication
- Management and information systems—these include the computer applications used to collect, synthesize, and analyze data to produce information and distribute that information to employees

Comparison of Affirmative Action, Managing Diversity and Valuing Differences

Affirmative Action	Managing Diversity	Valuing Differences
Quantitative. Emphasis is on achieving equality of opportunity in the work environment through the changing of organizational demographics. Progress is monitored by statistical reports and analyses.	**Behavioral.** Emphasis is on building specific skills and creating policies that get the best from every employee. Efforts are monitored by progress toward achieving goals and objectives.	**Qualitative.** Emphasis is on the appreciation of differences and the creation of an environment in which everyone feels valued and accepted. Progress is monitored by organizational surveys focused on attitudes and perceptions.
Legally driven. Written plans and statistical goals for specific groups are utilized. Reports are mandated by EEO laws and consent decrees.	**Strategically driven.** Behaviors and policies are seen as contributing to organizational goals and objectives, such as profit and productivity, and are tied to rewards and results.	**Ethically driven.** Moral and ethical imperatives drive this culture change.
Remedial. Specific target groups benefit as past wrongs are remedied. Previously excluded groups have an advantage.	**Pragmatic.** The organization benefits: Morale, profits, and productivity increase.	**Idealistic.** Everyone benefits. Everyone feels valued and accepted in an inclusive environment.
Assimilation Model. Model assumes that groups brought into system will adapt to existing organizational norms.	**Synergy Model.** Model assumes that diverse groups will create new ways of working together effectively in a pluralistic environment.	**Diversity Model.** Model assumes that groups will retain their characteristics and share the organization as well as be shaped by it, creating a common set of values.
Opens doors. Efforts affect hiring and promotion decisions in the organization.	**Opens the system.** Efforts affect managerial practices and policies.	**Opens attitudes, minds, and the culture.** Efforts affect attitudes of employees.
Resistance. Resistance is due to perceived limits to autonomy in decision making and perceived fears of reverse discrimination.	**Resistance.** Resistance is due to denial of demographic realities, the need for alternative approaches, and the benefits of change. It also arises from the difficulty of learning new skills, altering existing systems, and finding the time to work toward synergistic solutions.	**Resistance.** Resistance is due to a fear of change, discomfort with differences, and a desire to return to the "good old days."

Adapted from Lee Gardenswartz and Anita Rowe, *Managing Diversity: A Complete Desk Reference and Planning Guide*, 1993.

According to Craig (1993), "there's no magic formula for creating the 'right' workplace environment" (p. 18). Successful organizations share these qualities:

- A clear, shared vision which embodies positive values and drives people's behavior
- Leadership which communicates and reinforces the values
- Organizational members must be a valued asset and control their environment
- Organization must be adaptable and posses mechanisms that allow members to respond quickly and positively to a changing environment
- Improvement should be made based on measurements which reinforce shared values

SUMMARY

"If you need an expert on what it takes to get ahead in a U.S. organization if you aren't a white male, try Ann Morrison, president of New Leaders Institute," wrote Galagan (1993a, p. 39). Her research on barriers that hold women back led to her 1987 book, *Breaking the Glass Ceiling*. Her latest book, *The New Leaders: Guidelines on Leadership Diversity in America*, was the result of studying 16 model organizations. These organizations included 12 private-sector businesses, 2 government agencies, and 2 educational institutions. Some of these were American Express Company, Colgate-Palmolive Company, DuPont, Fairfax County (VA) Public Schools, Gannett, Kaiser Permanente, Michigan Bell, Motorola, the Palo Alto (CA) Police Department, U.S. West, and Xerox Corporation.

According to Morrison, in an interview by Galagan (1993a, p. 40), "unless the responsibility for advancement is shared, even women who make themselves into Superwomen won't be accepted in some organizations." Morrison found that "the single biggest barrier to advancement is prejudice—equating a difference with a deficiency" (p. 40). The next five barriers to advancement were poor career planning; a lonely, hostile, unsupportive working environment; lack of organizational savvy; greater comfort in dealing with one's own kind; and difficulty in balancing family and career. These factors and others are the direct result of learning and work environments as well as organizational cultures that do not value women and minorities nor understand the important contributions made by women and minorities.

Within the technology education profession, we must take action to discover and rediscover problems in our learning and work environments, strengthen our commitment to women and minorities in technology education, choose solutions that fit a balanced strategy, demand results and revisit regularly our goals and plans, and use successful approaches and achievements to maintain momentum.

Conceptually and demonstratively, we must work through a model of diversity that moves us well past an emphasis on compliance and management to a level where valuing diversity is prominent in the thinking and action of technology education professionals. Then we will ensure individuals from underrepresented groups will feel comfortable with their learning and work environment. Thus, they will feel more confident in their ability to contribute. When people contribute, they are more productive, and the environment within the profession will enable us to become a high-performance organization. As Margaret Mead noted, "If we are to achieve a rich culture, rich in contrasting values, we must recognize the whole gamut of human potentialities and so weave a less arbitrary social fabric, one in which each diverse human gift will find a fitting place."

REFERENCES

Anderson, J. A. (1993, April). Thinking about diversity. *Training & Development, 47*, 59-60.

Associated Press. (1996, October 1). Women's climb to top of business ladder slow. *The Pantagraph*, pp. C1, C4.

Carr, C. (1994, February). 7 keys to successful change. *Training*, 55-58, 60.

Craig, P. P. (1993). Pushing the right levers—The right way. *Journal of Business Strategy, 14*, 16-20.

Deal, T. E., & Kennedy, A. A. (1982). *Corporate culture: Rites and rituals of corporate life.* Reading, MA: Addison-Wesley.

Delatte, A. P., & Baytos, L. (1993, January). Guidelines for successful diversity training. *Training*, 55-56, 58-60.

Dovidio, J. (1993, April). The subtlety of racism. *Training and Development*, 51-57.

Ely, R. J. (1994). The effects of organizational demographics and social identity on relationships among professional women. *Administrative Science Quarterly, 39*, 203-238.

Galagan, P. A. (1993a, April). Diversity. *Training & Development*, 39-43.

Galagan, P. A. (1993b, April). Navigating the differences. *Training & Development, 47*, 29-33.

Galagan, P. A. (1993c, April). Trading places at Monsanto. *Training & Development, 47*, 45-49.

Goldstein, J., & Leopold, M. (1990). Corporate culture vs. ethnic culture. *Personnel Journal, 69*, 83-92.

Higgins, J. M. (1994). *The management challenge.* New York: Macmillan.

Hughes, R. (1992, February). The fraying of America. *Time, 31*, 44-49.

Jackson, B. W., Hardiman, R., & Chesler, M. (1981). *Racial awareness development in organizations.*

Jackson, B. W., LaFasto, F., Schultz, H. G., & Kelly, D. (1992, Spring/Summer). Introduction: Diversity, an old issue with a new face. *Human Resource Management, 31*(1 & 2), 21-34.

Kanter, R. M. (1977). *Men and women of the organization.* New York: Basic Books.

Kilmann, R. H. (1985, April). Corporate culture. *Psychology Today,* 62-65.

Knight-Ridder News Service. (1994, April 3). Women leave careers in science, engineering. *The Pantagraph,* p. C2.

Meares, L. B. (1986, July). A model for changing organizational culture. *Personnel,* 63, 38-42.

Porter, B. L., & Parker, W. S. (1992, Spring/Summer). Culture change. *Human Resource Management,* 31(1 & 2), 45-67.

Ridgeway, C. L. (1988). Gender differences in task groups: A status and legitimacy account. In M. Webster & M. Foschi (Eds.), *Status generalization: New theory and research.* Stanford, CA: Stanford University Press.

Rossett, A., & Bickham, T. (1994, January). Diversity training: Hope, faith, and cynicism. *Training,* 41-42, 43-45.

Strugatch, W. (1990, September). You & co: If you and your company have matching personalities, you'll get ahead—fast. *Self,* 206-210.

Thomas, R. R. (1990, March-April). From affirmative action to affirming diversity. *Harvard Business Review,* 107-117.

Van Eron, A. M. (August, 1995). Ways to assess diversity success. *HRMagazine,* 51-52.

 # Diversity in Technology Education

SUMMARY

Janet L. Robb
Gable Social Mobilization Campaign

IT IS TIME FOR THE PREACHERS, THE RABBIS, THE PRIESTS AND PUNDITS, AND THE PROFESSORS TO BELIEVE IN THE AWESOME WONDER OF DIVERSITY SO THAT THEY CAN TEACH THOSE WHO FOLLOW THEM.
(Angelou, 1993, p. 124)

People and education. My grandfather always advocated having faith in people and striving to become as educated as possible in all facets of life. My grandfather was one of the key people who influenced the course of my life, and I often find myself reflecting on his advice and searching for its inner meaning. His philosophy compels me to echo the words of Tom Peters (1994): "Personally, I find the polyglot parade of races and nationalities stimulating, fascinating, exciting. It's called diversity. It's a fact of life . . . It has long been my belief that diversity is an awesome opportunity as opposed to a problem" (pp. 219, 221).

Perhaps because of this philosophy, I was somewhat confused and overtly negative in 1984 when a fellow technology educator approached me to form a women's network within the profession. I know now that my negative response stemmed from the fact that I just didn't get it. In my desire to believe that everyone would be accepted and treated as equal if he or she just persevered (and conformed), I failed to recognize the value of support and encouragement by those facing similar obstacles and similar successes.

As practitioners read the chapters in this yearbook, they will most likely be struck by the tenor that illustrates that, as professionals, we have not encouraged, assisted—or found as commonplace—diversity. When I was active as a technology educator, I thought I was advocating that if one could live life with a commitment to equity, then others would recognize him or her as an equal. But what I was really doing was saying, "Conform. Don't be unique. Be like everyone else in the 'good old boy system' and you won't need any other network but the good old boys." I have since learned that without support and advocacy, being a female technology educator became a missed opportunity to share the characteristics unique to gender, ethnicity, and background. Had I remained in the profession, I might well have never been accepted for who I was, but for who I was willing to be.

I failed to get the "in all facets of life" part of my grandfather's philosophy. As I reflect, I realize that my grandfather was not only advocating more diplomas and more degrees, but he was saying that people need to be educated, be it about school subjects or social structures, before they can truly understand, accept, and embrace an issue, an ideal, or a way of life. As one young Hispanic female university student said about diversity in culture: "Knowing of other cultures doesn't necessarily mean we know them. Even though I'm aware of differing lifestyles, I don't understand them. I am, however, beginning to realize that understanding my neighbor's culture can only enrich my own life" (Larsen-Pusey, 1993, p. 137). I believe this is not only true of culture, but of all forms of diversity. The only way we can achieve the goal of inclusion and acceptance is through education. As Headlam (1993) wrote,

> We need to develop an intellectual and emotional understanding of each other so that we can finally come to value our diversity and realize the full potential of what we have in human resources. We need an education of both our minds and our hearts to unlock the barriers that separate us, rich and poor, White and Black, men and women. (p. 10)

TECHNOLOGY EDUCATION'S ROLE

Granted, diversity is not a new issue. But traditionally, in our schools, the business world, and society at large, our approach to diversity has been assimilation. Newcomers are expected to "fit in," and the burden of adjustment falls to them. In the past, most newcomers bought into the rationale being successful required uniformity and that meant conforming. Gollnick and Chinn (1994) summarized this concept:

> A popular assimilation theory is what is called the *melting pot*. . . . It appealed to many immigrants, especially those with European backgrounds. It predicted the evolution of a new, unique American culture to which all ethnic groups contribute. This theory captured an egalitarian ideal, but in reality many groups were not allowed to melt because of the nation's racist policies and practices. (p. 17)

In technology education, whether at secondary or tertiary levels, as endorsed through the chapters presented in this yearbook, the acceptable practice has been to expect the newcomer to adjust to a well-established system. "In anticipation of promised success as a result of conforming, they (the newcomers) have dropped their ethnic and gender identity" (Thomas, 1991, p. 8).

For all the conforming and assimilating, the attraction and retention of diverse groups in technology education is not apparent. There still appears to be a dearth of minorities and women at all levels of the technology education spectrum. This inequity implies that a new approach may be needed. As Liedtke, in her chapter, wrote of the young African-American girl who was captivated by lasers and robotics, "the educational environment must change to accommodate her." As Thomas (1991) reiterated,

> People being assimilated are never quite comfortable. They find themselves caught between two worlds and uncomfortable in both. What is perhaps more damaging, forcing everyone to assimilate leaves untapped potential. Because assimilating people want to fit in, they focus on doing the expected or accommodating the norm, on playing it safe. They avoid offering suggestions that would make them stand out. They can't focus on their personal strengths or on innovative ideas; they're too busy trying to adapt. The consequences can be a lackluster performance. In a competitive environment, assimilation is stifling and deadly. (pp. 8-9)

Diversity in Technology Education

Technology education advocates diversity in its textbooks, classrooms, and laboratories—diversity in materials, techniques, processes, problems, and solutions. As technology educators, we encourage our students to learn everything about the materials they are going to use: their chemical makeup, how they react under different stimuli, etc. We advocate exploration and scientific methodology, and we promote creativity and diversity in problem solving. If we can adapt this approach to human inputs in our classrooms and profession, I believe we might just make great strides in embracing diversity among our students, teachers, and teacher educators.

Allen and Hutchinson (1994, p. 347) reported that "few teacher education programs include information about diverse cultures and less still about teaching strategies and techniques to enhance the success and self-confidence of ALL students." We must be willing to learn everything we can about the diversity we are encountering, be it gender, ethnicity, ageism, disability, or a variety of other characteristics that create diversity among us—social class, sexual orientation, religion, language, etc. We must find ways to accept and encourage a heterogeneous population, whether in our classrooms or profession.

To this end, I leave you with four positive steps that need to be taken and three dangers to be avoided.

Steps To Be Taken

1. Infuse the principles and practices of multicultural education throughout your courses. Do not leave the task to isolated courses with such titles as "Teaching the Culturally Different," "Handling Diversity," "Bilingual Education," etc. (Larken & Sleeter, 1995). "Multicultural education incorporates the idea that all students, regardless of their social-class, racial, ethnic, or gender characteristics, should have an equal opportunity to learn" (Banks, 1994, p. 10).

2. Adopt a new metaphor for diversity. "A symphony, a mosaic, a salad bowl are all metaphors that quickly come to mind; these more accurately describe the social dynamics occurring

among diverse groups in this country than the old 'melting pot' metaphor" (McCormick, 1994, p. 45). These new metaphors create images that require a variety of different pieces to make a whole. People perform better, feel more comfortable, and become major contributors when they feel as though they matter and their presence is important.

3. Be a culturally responsive educator and professional. View each student and each colleague as unique. According to Lee (1994, p. 32), "Culturally responsive educators: Challenge their own prejudicial thinking about culturally diverse students. Seek an understanding of cultural diversity. Integrate the accomplishments of diverse cultures into existing curriculum. Involve culturally diverse parents and community resources in schools." With only slight modification, these characteristics can be used to define culturally responsive professionals as those who challenge their prejudicial thinking about culturally diverse colleagues, seek to understand diversity, assure that the accomplishments of diverse groups are recognized and rewarded, and actively involve diverse groups, organizations, and individuals.

4. Passionately embrace diversity. "Educating for diversity requires passion: passion for people, their cultures and languages; passion for human potential; passion for the process and products of change" (Rios & Whitehorse, 1994, p. 37). A passionate approach to diversity accepts and promotes networks and support groups and contributes to their effectiveness.

Dangers to Avoid

1. Collision with the rhetoric. Using the correct words and going through the proper motions, without an underlying acceptance of the uniqueness and value of diversity, only allows inequities and alienation to continue and to grow.

2. The assumption that diversity will go away. All reports indicate that by 2000, one of every three Americans will be non-White.

As Barnette pointed out in the chapter, "Minority Students," 39% of the school population will be made up of minority students by 2020. Couple this with the reality that more women and disabled are entering the workforce every year and more people are acknowledging other diversities, such as sexual orientation, age, language, and religion. Assuming that diversity will go away hinders the profession from reaching its maximal potential on all levels—intellectually, politically, morally, and ethically.

3. Overzealous passion. Not everyone has been convinced that the best way to approach diversity is through acceptance and understanding. Many still favor the melting pot approach. They feel they had to conform, and they survived, so everyone else should conform, as well. It will take time for some people to recognize the value in celebrating diversity, as opposed to ignoring it. As Rios and Whitehorse (1994, p. 37) stated, "unbridled passion, exemplified in zealousness, can be as harmful as apathy."

Recognizing, exploring, understanding, and accepting diversity will make for a healthier, happier classroom and professional environment. It will make everyone feel a part of the whole and allow the entire profession to recognize its full potential. It will allow all to be accepted and appreciated for who they are, not for who they are willing to become. As Angelou explained, "We all should know that diversity makes for a rich tapestry, and we must understand that all the threads of the tapestry are equal in value no matter their color; equal in importance no matter their texture" (1993, p. 124).

REFERENCES

Angelou, M. (1993). *Wouldn't take nothing for my journey now.* New York: Random House.

Allen, K. W., & Hutchinson, C. J. (1994). Education reform and issues of diversity. In C. A. Grant (Ed.), *1994 National Association for Multicultural Education Proceedings* (pp. 347-356). Los Angeles: Simon & Schuster Elementary Education Group.

Banks, J. A. (1994). Dimensions of multicultural education. *Insight on Diversity,* 10.

Gollnick, D. M., & Chinn, P. C. (1994). *Multicultural education in a pluralistic society* (4th ed.). New York: Macmillan.

Headlam, A. (1993). Under the bridge: My turn. In C. A. Grant (Ed.), *1993 National Association for Multicultural Education Proceedings* (pp. 9-12). Los Angeles: Simon & Schuster Elementary Education Group.

Larken, J. M., & Sleeter, C. E. (1995). *Developing multicultural teacher education curricula.* Albany: State University of New York Press.

Larsen-Pusey, M. A. (1993). Case studies and learning logs: Celebrating diversity. In C. A. Grant (Ed.), *1993 National Association for Multicultural Education Proceedings* (pp. 133-140). Los Angeles: Simon & Schuster Elementary Education Group.

Lee, C. C. (1994). Being culturally responsive. *Insight on Diversity,* 32.

Liedtke, J. A. (1998). Environmental and climate challenges in technology education. In B. L. Rider (Ed.), *Diversity in Technology Education.* New York: Glencoe

McCormick, T. E. (1994). Reverence for human diversity. *Insight on Diversity,* 45.

Peters, T. (1994). *The pursuit of WOW! Every person's guide to topsy-turvy times.* New York: Vintage Books.

Rios, F. A., & Whitehorse, D. M. (1994). A passion for multicultural education. *Insight on Diversity,* 37.

Thomas, R. R., Jr. (1991). *Beyond race and gender.* New York: American Management Association.

INDEX

A

Abolitionist movement, 15
Academy, 160
Addams, Jane, 24
Affirmative action, 168–69, 173
Africa
 origin of human race in, 42
 population in, 2
African-Americans
 contributions of, to origin and evolution of various technologies, 41–42
 contributions of, to technology, 47
 contributions of, to technology education, 37–51
 early involvement of, in industrial education, 38–40
 learning styles of, 102–3
 prospectus on technology education, 50–51
Agenda, setting, in mentoring relationship, 136
Aggression, gender differences in, 105
Albright, Madeline, 142
American Council for Elementary School Industrial Arts (ACESIA), 27
American Council on Industrial Arts Teacher Education (ACIATE) Yearbook, 26
American Industrial Arts Association (AIAA), women's roles in, 25
American Institute for Managing Diversity, 151
Analytical learners, 114
Analytical style, 108
Anthony, Susan B., 14
Armstrong, Samuel Chapman, 39–40
Asia, population in, 2
Assimilation theory, 181
Attila, 141
Attrition in American colleges and universities, 85–86
Australia, population in, 2

B

Banneker, Benjamin, 48
Baseball Team, 160

Beard, Andrew Jackson, 49
Beer-manufacturing technology, 44
Behavioral norms, 157, 158
Bet-your-company culture, 159
Biological clock, 17
Bio-related technology, 44
Blackburn, A. B., 49
Boardman, Alice, 21
Bonser, Frederick Gordon, 13, 26
BOWGSAT method, 165
Boyd, Henry, 48
Brodie, James Michael, 47
Bush, George, 151
Business issues, interest in, 58

C

Canada, population in, 2
Capital mobility, 4
Career induction, 155
Caregivers, learning styles of, 100
Carver, George Washington, 49
Catherine the Great, 142
Ceramics technology, 45
Ceremonies, 157, 158
Certification of minority students, 90–91
Chambers of Commerce, 145
Change as inevitable, 1

Charismatic leaders, 141–42
  effectiveness of, 142–43
Civitan, 145
Club, 160, 168
Coffey, Lois Mossman, 26
Cognitive abilities, gender differences in, 105
Communication skills, 58
  as leadership tenet, 144–45
Competitive learning environment, 101
Compliance, 165, 168
Computers
  and data analysis, 115
  in supporting diverse learning styles, 114–15
Conducive learning environment, 102–3
Conformity, 160
Construction technology, 43–44
Conversational style, 106
Cooperative learning activities, 113–14
  positive intergroup outcomes from, 114
Cooperative learning environments, 101
Corporate culture, kinds of, 159–64
Council on Technology Teacher Education (CTTE), iv

Cultural learning styles, 101, 102
  of African-Americans, 102–3
  analytical, 108
  differences in, 100–101
  effect of culture on, 99
  of European Americans, 103–4
  of female students, 105–6
  of Latin-Hispanic Americans, 103
  of Native Americans, 104
  relational, 108
Culturally responsive educators, 183
Cultures
  creating change, 164–74
  definition of, 100
  effect of, on learning styles, 99
Curriculum
  delivery of, 100–101
  differences in, 100–101

D

Decision making as leadership tenet, 143–44
Demographics
  changing, of organization, 164
  of minority students, 78–79
  world, 1–7
Developed countries, effect of developing countries on, 3
Developing countries, effect on developed countries, 3
Dewey, J., 24
Diana (Princess of Wales), 142
Dictatorial leaders, 142, 144
  effectiveness of, 142–43
Discrimination, gender, 59–60
Discussions, 103
Diversity, 164–65
  adopting metaphor for, 182–83
  of minority students, 89–90
  programs in, 151
  steps for managing cultural, 165–66
  teaching strategies for, 113–15
  in technology education, 1–8, 179–84
  in United States, 97–98
  view of, as problem, 97
  of world population, 4–6
Dodge, Grace, 13, 20
Dole, Elizabeth, 142
Domestic arts, 21
Downing, P. B., 49
Drew, Charles Richard, 49
Dubois, W. E. B., 40

E

Eastern Michigan University, creation of Department of Industrial Arts at, 21
Educational attainment, 160
Educational environment, confluence of change and diversity in, iii
Egypt
 bio-related technology in, 44
 ceramics technology in, 45
 construction technology in, 43–44
 manufacturing technology in, 44
 metallurgy technology in, 46
 pyramids in, 42
 transportation technology in, 45–46
Elementary school
 industrial education and manual training programs in, 19
 influence on industrial movement, 20–21
Employment, gender and racial inequality in, 6
Environmental and climate challenges in technology education, 151–75
Environmental learning style, 103
Equal rights amendment, 17
European American learning styles, 103–4
Exclusionary organization, 167

F

Face-to-face promotive interaction, 114
Family/outside pressures on women, 60–63
Feedback in mentoring relationship, 133
Feminist movement, 15
 growth of, 13–14
Fermentation, 44
Fortress, 160
Freedman's Bureau, 39

G

Gender
 differences in mathematical ability, 105
 and discrimination and stereotyping, 59–60
 finding equitable solutions to inequitable situations, 69–70
 in global society, 5
 and learning styles, 100–101
Gender-fair language, 111–13
Gender-neutral language, 111
Glass ceiling, 152

Glass walls, 152
Global economy, development of, 4
Global society
  effect of migration on, 3
  gender in, 5
Good old boy system, 130, 180
Graduate programs
  importance of mentoring during, 125–26
  in technology education, 50–51
Group processing, 114

H

Hampton Institute, 39, 40
Harmony, 165
Heasley, Norma, 26, 27
Hennes, M., 21
Hierarchy in mentoring relationship, 136
High-risk students, profile of, 85
High school, technology education in, 106
Hilo Manual Labor School for Native Hawaiians, 39
Hispanics, 98
  learning styles of, 103
Historically Black Colleges and Universities (HBCUs), 50, 84
Hitler, Adolf, 142
Holistic learners, 114
Home economics education, influence on industrial movement, 20–21
Humor as leadership tenet, 145
Hunt, Elizabeth, 26
Huntington, Emily, 13, 20

I

Iacoca, Lee, 142
Identifiable cultures, 159–64
Identifiable value systems, 157
Image, increasing positive, through language, 111–13
Inclusion, 165
Individual accountability, 114
Individual defect paradigm, 66–67
Individuality, 162
Individual societies, technological effects of world on, 1
Industrial arts, 24
  women's contributions to, 16–17, 21–25
Industrial Arts for Elementary Schools, 23
Industrial education
  definition of, 23–24
  involvement of Africans and African-Americans in, 38–40

women's contributions to, 14–15, 19–21
Industrial Education Association, 20, 28
In-school meetings, 109
Instinct, 160
Interactions, 103
Intercultural relations
  concerns about, 98–99
  cooperative learning as mechanism in improving, 113–14
  quality of, in schools and classrooms, 107
International Technology Education Association (ITEA), 27, 51
  leadership roles in, 77
  mission statement in its strategic plan, 6–7
  strategic planning goals of, 64
International trade, 4
Interpersonal skills, 114

J

Jefferson, Thomas, 41–42
Jennings, Thomas L., 48
Jobs, relocation of, 4
Justice, 165

K

K-12 technology education programs, profiles of minority students in, 80–82
Kinesthetic learners, 103, 115

Kitchen Garden Association, 20, 28
Kiwanis, 145

L

Language, increasing positive image through, 111–13
Language systems, 157, 158
Lathe, 44
Latimer, Lewis Howard, 48
Latin America, population in, 2
Latin/Hispanic American learning styles, 103
Leadership
  communication in, 144–45
  decision making in, 143–44
  effectiveness of styles, 142–43
  humor in, 145
  research on, 141
  risk taking in, 146–47
  roles in International Technology Education Association (ITEA), 77
  styles of, 141–43
Learning. See also Cultural learning styles
  diverse teaching strategies for diverse, 113–15
  role of teachers in, 107–10

Learning environment
  creating suitable, 99
  technology in, 153
Lee, Joseph, 49
Legitimacy, women's struggle for, 57–58
Limited English proficiency (LEP), 99
Listening skills, 58, 144
  in mentoring relationship, 133
Literacy
  technological, 3, 57
  of women, 2
Lockette, Rutherford E., Humanitarian Award, 51
Long-term commitment, 160
Loom, 44

M

Majority minority population, 98
Male learning patterns in classrooms, 106
Management and information processes, 172
Management and information systems, 172
Managing diversity, 173
Manual training, 21
Manufacturing
  role of women in, 15
  technology in, 44
Masonic Lodge, 145
Mathematical ability, gender differences in, 105
Matzeliger, Jan, 48
McCoy, Elijah, 48
Mead, Margaret, 57
Melting pot metaphor, 4, 181, 183
Menes, 43
Mentoring, 64
  benefits (and risks) to individuals involved, 135–36
  contrast with social networking, 135–36
  determining time, for ending, 133
  guidelines for, 131–35
  hierarchy in, 136
  importance of, during graduate study, 125–26
  of minority students, 87–89
  myths about, 129–30
  suggestions for future, 137–38
  women in technology, 126–27, 130–31
  women in workplace, 127–29
Mentors
  appropriateness of relationship with protégée, 127
  in recruiting women in technology education, 65
  selection of, 130
Mentors of moment (MOMs), 131

Index

Metallurgy technology, 46
Metaphors, 157
  adopting for diversity, 182–83
  and language systems, 158
  melting pot, 4, 181, 183
  mosaic, 4, 66–69
  salad bowl, 4
Middle school, technology education in, 106
Migration, effect of global society on, 3
Minority population, increase in United States, 5
Minority students, 77–91
  certification of, 90–91
  demographics of, 78–79
  diversity in, 89–90
  identifying, 78
  in K-12 technology education programs, 80–82
  mentoring of, 87–89
  mentors for, 87
  in post-secondary programs, 82–85
  recruitment and retention of, 85–86
Mississippi Valley Industrial Teacher Education Conference, 27–28
Monroe, John, 51
Montessori, Maria, 30
Morgna, Garrett Augustus, 49
Morrill Act, 40

Mosaic metaphor, 4, 66–69
Mossman, Lois Coffey, 13
Mother Teresa, 142
Multicultural education
  infusing principles and practices in, 182
  role of teachers in, 110
Multiculturalism, National Science Teachers Association (NSTA) policy statement on, 102
Multicultural organization, 170–72, 174
  principles for identifying progress toward, 166–67
Myths, 157
  about mentoring, 129–30
  about women, 151–52

N

National Academy of Sciences, 137
National Aeronautics and Space Administration (NASA), 51
National Education Association's National Center for Innovation, 104
National Science Foundation (NSF), 51
National Science Teachers Association (NSTA), policy statement on multiculturalism, 102
National Teacher Exam, 90

Native American learning styles, 104
Native intelligence, 58
Nesting urge, 17
Networking, 126–27
    in recruiting women in technology education, 65
New Zealand, population in, 2
Nontraditional professions, women in, 15

O

Oceania, population in, 2
Ohio State University Elementary School and Kindergarten, 24
Old-Boy Network, 125–26
Organizational culture, 156–58
    definition of, 156
    increasing participation of women and minorities in, 156
Organization structure, 172

P

Patton, George, 142
Penney, J.C., female role models at, 129
Personal training, value of, 160
Person with disability, individual rights of, 67
Philadelphia Women's Network, 129
Pink ghetto, 18
Pinney, 20–21
Political skills, 58
Population
    distribution of, 2
    size of, 2
Positive interdependence, 114
Post-secondary programs, profiles of minority students in, 82–85
Potentials, 152
Praxis, 90
Prevocationalism, 22
Process culture, 159–60
Professional organizations, role of women in, 28
Protégées. See also Mentoring; Mentors
    appropriateness of relationship with mentors, 127
Psychological learning style, 103
Purchasing agents, 143

Q

Quality of life, factors effecting, 3
Queen bee syndrome, 162

R

Racial diversity, 4–5
Racial issues in workplace, 151
Reading instruction, 97–116
    and cultural learning styles, 102–6

diverse teaching strategies for diverse learners, 113–15
and factors influencing participation in technology education, 106–7
increasing positive image through language, 111–13
role of teachers in learning, 107–10
Reagan, Ronald, 151
Recruitment and retention
of minority students, 85–86
of underrepresented groups, 161
of women, 63–69
Redefining organization, 169
Relational style, 108
Relocation of jobs, 4
Rillieux, Norbert, 48
Risk taking as leadership tenet, 146–47
Rituals, 157, 158
Role models, 64–65
Roosevelt, Franklin Delano, 142
Rotary, 145

S

Sagas, 157
Salad bowl metaphor, 4
Salience and mentoring, 88
School-aged population, multicultural, 99
Scientific management, 104
Scobey, Mary Margaret, 26, 28–29
Secondary vocational education programs, 80–81
Segregation, 50
Self-confidence/assertiveness, 58
Self-discovery, 114–15
Self-esteem, enhancing, 113
Sensitivity, 58
Service clubs, 145
Skill training, 22
Slaves, skill level of, 38
Small-group skills, 114
Smelting, 46
Social distance and mentoring, 88
Social networking, mentoring contrast with, 135–36
Social reconstruction, 25
Society
expectations of women in, 13–14
and technology education, 1–8
Soldering, 46
Spatial ability, gender differences in, 105
Specialization, 160
Stanton, Elizabeth Cady, 14

STEM (science, technology, engineering, and mathematics) mold, 127
Step-by-step program procedures, 114
Stereotyping, 59–60
Success, achieving, 58–59
Suffrage, 15, 16
　universal, 14
Sumaria, 46
Summer camps in recruiting women in technology education, 65
Support/encouragement, effect of, on women, 60–63
Symbols, 157, 158

T

Teachers
　in creating multicultural environment, 110
　culturally responsive, 183
　monocultural, 99
　role of, in learning, 107–10
　traditional methods of, 109
　use of diverse strategies for diverse learners, 113–15
　women as, 15
Teachers College Columbia, 20
Teachers College Record, women's contributions to, 21–22
Teaching Children About Technology, 26
Teaching of Industrial Arts in Elementary School, 23
A Technological Exploratorium, 26–27
Technological literacy, 3, 57
Technology
　African-Americans contributions to, 41–42, 47
　bio-related, 44
　ceramics, 45
　construction, 43–44
　manufacturing, 44
　mentoring women in, 130–31
　metallurgy, 46
　transportation, 45–46
Technology education
　certification in overcoming barriers in, 90–91
　challenges in, 153–56
　changes in, 6–7
　contributions of African-Americans to, 37–51
　contributions of women to professional literature in, 28–29
　in creating culture change, 164–74
　diversity in, 179–84

diversity in overcoming
barriers in, 89–90
environmental and climate challenges in,
151–75
factors influencing participation in, 106–7
historical view of
women's roles in,
13–31
identifiable cultures in,
159–64
increasing enrollment of
women in, 153
during industrial arts era,
16–17
during industrial education era, 14–15
mentoring in overcoming
barriers in, 87–89
opportunities for, 3
organizational culture of,
156–58
recruitment and retention
in overcoming barriers in, 64–69, 85–86
society and diversity in,
1–8
during technology education era, 17–18
women's contributions to
evolution of, 19–31
Technology Education
Collegiate Association
(TECA), 84, 85
Technology Education for
Children Council, 27

Technology educators,
women as, 57–70
Technology for All
Americans Project, 6,
7
Technology for Children
(T4C), 26
Technology learning environment, 153
Technology profession,
mentoring women in,
126–27
Technology Student
Association (TSA), 78,
84, 85
Tech Prep model, 83
Tough-guy/macho culture,
159
Transactional leader, 142,
144
Transformation, 165,
171–72
Transformational leaders,
142, 145
Transportation technology,
45–46
Trust and mentoring, 88
Tuskegee Institute, 40
2+2 model, 83

U

Underrepresented groups,
161–64
technology teachers
from, 155–56

United States
  concerns about intercultural relations in, 98–99
  diversity in, 97–98
  minority population increase in, 5
  population in, 2
Universal suffrage, 14
University laboratory schools, 24
U.S.S.R., population in, 2

V

Value systems, 158
Valuing differences, 173
Verbal ability, gender differences in, 105

W

Washington, Booker T., 40
White population, 98
Whitney, Eli, 41
Women
  in achieving success, 58–59
  in American Industrial Arts Association (AIAA), 25
  contributions to professional literature, 28–29
  factors influencing participation in technology education, 106–7
  finding equitable solutions to gender-inequitable situations, 69–70
  and gender discrimination and stereotyping, 59–60
  historical view of roles in technology education, 13–31
  increasing enrollment of, in technology education, 153
  in industrial arts, 16–17, 21–25
  in industrial education, 14–15, 19–21
  learning styles of, 105–6
  literacy of, 2
  mentoring
    benefits and risks of, 135–36
    during graduate study, 125–26
    guidelines for, 131–35
    myths about, 129–30
    suggestions for, in the future, 137–38
    in technology, 126–27, 130–31
    in workplace, 127–29
  myths about, 151–52
  in professional organizations, 28
  recruitment, retention and advancement of, 63–69

societal expectations of, 13–14
struggle for legitimacy, 57–58
support/encouragement and family/outside pressures on, 60–63
in technology education, 17–18, 26–28 as technology educators, 57–70
turnover rate for high-potential, 162–63
in workplace, 2–3, 97
in world population, 2
Woods, Granville T., 48
Woolman, 20–21
Work hard/play hard culture, 159
Workplace
 importance of mentoring women in, 127–29
 racial issues in, 151
 women in, 2–3, 97

Work processes, 172
Workshops, 109
World demographics, 1–7
 literacy of women, 2
World population
 diversity of, 4–6
 women in, 2
World War II, women in, 15
Writing instruction, 97–116
 and cultural learning styles, 102–6
 diverse teaching strategies for diverse learners, 113–15
 and factors influencing participation in technology education, 106–7
 increasing positive image through language, 111–13
 role of teachers in learning, 107–10